# Geology for Nongeologists

FRANK R. SPELLMAN

Government Institutes
An imprint of
The Scarecrow Press, Inc.
Lanham, Maryland • Toronto • Plymouth, UK
2009

**Government Institutes**

Published in the United States of America
by Government Institutes, an imprint of The Scarecrow Press, Inc.
A wholly owned subsidiary of
The Rowman & Littlefield Publishing Group, Inc.
4501 Forbes Boulevard, Suite 200
Lanham, Maryland 20706
http://www.govinstpress.com/

Estover Road
Plymouth PL6 7PY
United Kingdom

The reader should not rely on this publication to address specific questions that apply to a particular set of facts. The author and the publisher make no representation or warranty, express or implied, as to the completeness, correctness, or utility of the information in this publication. In addition, the author and the publisher assume no liability of any kind whatsoever resulting from the use of or reliance upon the contents of this book.

British Library Cataloguing in Publication Information Available

**Library of Congress Cataloging-in-Publication Data**

Spellman, Frank R.
   Geology for nongeologists / Frank R. Spellman.
     p.  cm.
   Includes bibliographical references and index.
   ISBN 978-0-86587-185-4 (pbk. : alk. paper)—ISBN 978-1-59191-944-5 (electronic)
   1. Geology—Popular works.   I. Title.
   QE31.S64   2008
   550—dc22
                                            2008055604

♾ ™ The paper used in this publication meets the minimum requirements of American National Standard for Information Sciences—Permanence of Paper for Printed Library Materials, ANSI/NISO Z39.48-1992. Manufactured in the United States of America.

For
Kimberly Ann Barnes

Under heaven nothing is more soft and yielding than water.
Yet for attacking the solid and strong, nothing is better;
It has no equal.

—Lao Tzu, *Tao Te Ching*

# Contents

# Preface

The purpose of this book is to present basic concepts in geology, with the focus primarily on how rocks, minerals, and fossils are classified; how wind, ice, and water have shaped the earth; how mountains are formed; as well as how volcanoes, geysers, earthquakes, glaciers, and groundwater work to modify the physical structure of Earth. In so doing, this book fills the gap between general introductory science texts and the more advanced environmental science books used in graduate courses. Thus, *Geology for Nongeologists* presents a nontechnical survey of geology in reader-friendly written style and provides a levelheaded look at a very serious discipline, one based on the geological record and years of extensive research.

Geology is a multidisciplinary field that incorporates aspects of biology, chemistry, physics, ecology, geography, meteorology, pedology, and many other fields. Books on the subject are typically geared toward professionals in these fields. This makes undertaking a study of geology daunting for those without this specific background. However, this complexity also indicates geology's broad scope of impact. Because geologic phenomena affect us in such diverse ways (sometimes in profound ways such as volcanism and earthquakes), it is important to understand some basic concepts of the discipline.

This book teaches what everyone needs to know about geology and how science and scientists work. Readers will discover a new appreciation for their surroundings. Most importantly, it is hoped that this book will enable you to broaden your outlook on the environment we call Earth.

*Geology for Nongeologists*, again, is suitable for use by both the technical practitioner in the field and by students in the classroom; it emphasizes basic concepts, definitions, and descriptions, all blended with a touch of irony. The following list provides a broad-brush overview of the book, its contents, and how we proceed.

1. Study of geology can increase literacy in general science.
2. Science comes from observations.
3. The Earth is a planet.
4. The Earth is very old.
5. Planet Earth is a unique, evolving system.
6. Plate tectonics serves as a grand unifying theory to explain Earth processes.
7. Geological phenomena (earthquakes, volcanic eruptions, hurricanes, landslides) affect our environment.
8. The physical aspects of Earth systems are linked to life processes.

To ensure correlation to modern practice and design, illustrative problems are presented in terms of commonly used geological parameters. Here is all the information you need to make technical and personal decisions about geology. It is important to remember, however, that Mother Nature can perform wonders, but overload her and there might be hell to pay (e.g., earthquake or volcanic eruption).

Each chapter ends with a chapter review test to help you evaluate your mastery of the concepts presented. Before going on to the next chapter, you should take the review test, compare your answers to the key provided in the appendix, and review the pertinent information for any problems you missed. If you miss many items, review the whole chapter.

Again, this text makes the concepts of geology accessible to those who have no experience with the discipline. If you work through the text systematically, you will be surprised at how easily you acquire an understanding and skill in geology—adding a critical component to your professional knowledge.

Frank R. Spellman
Norfolk, Virginia

# CHAPTER 1

# Introduction

Nobody hurries geology.

—Mark Twain

*Earth*, *Planet Earth*, the *World*, and *Terra*, the third planet from the sun, the place we live—what could be more important to us? Yet, how little most of us really know about the composition and history of the planet that is our home.

—Jennifer Blue

The statements above set the stage for what is to follow in this text. As Mark Twain points out, geology is all about time—lots of time, eons of time, time that is endless (hopefully). The fact is—because of the passage of time—we do not know what we do not know about our own planet, Earth. How could we? None of us has lived the 4.54 billion years from the time the scientific evidence suggests that Earth was formed to the present. Well, we might think, to find out about Earth—its beginning and its subsequent history—couldn't we just refer back to the written record? Written record? Depends on what written record is being referred to. In regards to the human written record, keep in mind that prior to 5,000–5,500 years ago there is no written record (that we are aware of) that has preserved all the knowledge of facts or events that occurred in the past. There simply is not a continuous narrative and systematic analysis of past events of importance to the human race because on the scale of Earth's existence the time humankind has spent on Earth is measured as a small drop in an enormous ocean, and the period in which the ability to write exists is so small it cannot be perceptively measured. Events that occurred before the advent of written communication are dubbed "prehistory." In writing—the written historical record of Earth—there is little that is known and/or written about prehistoric times.

The written historical record is defined as written documents preserving knowledge of facts and events. As pointed out above, the written historical record as recorded by humankind is limited—extremely limited. This is not to say, however, that there is no record of the history of Earth available for our perusal. No one has written more about the history of past events of importance to Earth and the human race than Mother Nature. All we need do is to look at Earth (rocks) and its formations to see its past. It is true, however, that not all of what Mother Nature has written in Earth's past is that clear to us. The heavy hand of time has molded, shaped, reshaped, retooled, and reshaped again and again Earth's physical features in so many different ways and in so many directions that it is literally impossible for anyone to decipher exactly what has occurred to any given region on Earth at any given time in the ancient past. Compounding the problem brought about by the effects of time is that there are not that many of us around at present who are able or qualified to decipher what it is we are looking at in regards to Earth and its history. To do this to any extent, we must know geology.

Without thinking about it (without any great level of geological knowledge), there are some geological things we all know. We know about the products of soils that have been formed by weathered rock, oil formed from the remains of prehistoric plants and animals, and the beauty and value of precious stones. These are the basics—but only a small fraction of the useful materials with which Earth provides us.

Think about how impossible it would be for modern industry to have developed as we know it today without other Earth products. Mineral resources such as coal, iron, lead, and petroleum derived from the Earth have been made readily available through the application of basic geology, geological engineering, global positioning systems (GPS), and generic mapping tools (GMT).

Earth also provides us with areas of exceptional beauty. One need not be a geologist to marvel at the breathtaking vastness of the Grand Canyon; the remarkable sandstone formations and the very color of the earth at Zion, Bryce Canyon, the Arches National Park, and Monument Valley; the mystery of Luray Caverns; or the natural wonders of Yosemite or Yellowstone. All of these and many more, are the results of geologic processes that are dynamic (still at work today). They are the same geologic processes that began to shape Earth more than 4.54 billion years ago!

# What Is Geology?

It is important to point out that geology is not only the study of Earth as we see it today, but the history of Earth as it has evolved to its present condition. In attempting to pin down a definition of geology (to explain what it is and what it is about), consider the definition provided by Press and Siever (2001):

> Earth is a unique place, home to more than a million life forms, including ourselves. No other planet yet discovered has the same delicate balance necessary to sustain life. Geology is the science that studies the Earth—how it was born, how it evolved, how it works, and how we can help it.

## Did You Know?

- The earth has evolved (changed) throughout its history, and will continue to evolve.
- The earth is about 4.6 billion years old; human beings have been around for only the past 2 million years.

The term *geology* is derived from the Greek *geo*, meaning "earth," plus *logos*, "discourse or study of." Paraphrasing Press and Siever's (2001) definition of geology in simplistic terms, we can define it as the science that deals with the origin, structure, and history of Earth and its inhabitants as recorded in the rocks.

Geology consists of the sciences of mathematics, physics (geophysics and seismology), and chemistry (petrology and geochemistry); these are the sciences that set the general principles for all the other sciences. Additionally, geology is composed of or interrelated with the sciences that describe the great systems that make up the universe: astronomy (planetary geology) and biology (paleontology). Some geologists, the specialists, deal with very narrow and specialized parts of geology including mineralogy, meteorology, botany, zoology, and others. Geological science also includes exposure to the principles of sociology and psychology. In a nutshell and based on personal experience, it can be said with some certainty that the practicing geologist, whether a specialist or not, should and must be a generalist. That is, to be effective in practice, the geologist must have a well-rounded exposure to and some knowledge of just about everything related to the other scientific disciplines.

Professional geologists work to locate geologic resources. They also perform geological and mining engineering, site studies, and land-use planning. In regards to environmental protection, they conduct environmental impact studies and perform groundwater and waste management functions. Geologists also conduct basic research to maintain the science on the cutting edge of knowledge for current and future applications.

## The Scientific Method

We certainly cannot have a discussion about geology (or any other science) without a brief discussion of the scientific method of investigation. This is the case even though many scientists feel that there is no single universally accepted "scientific method" (see figure 1.1).

The step-by-step procedure used in the scientific method:

1. Observe and define the problem.
2. Hypothesize a theory (reasonable explanations based on observations).
3. Test (experiment on) the theory. (Can it be reproduced?)
4. Can others reproduce the same results?
5. If steps 3 and 4 are unsuccessful, modify or reject theory (go back to step 1).
6. If it is consistently reproducible and most of the scientific community agrees with the theory, it becomes a natural law.

Most scientific thought is based on two forms of logical reasoning: inductive and deductive reasoning. The goal of science is to establish general principles for the study of specific cases. The classic example is Newton's apple. Newton examined how specific objects fall, from which he inferred the general theory of gravitation.

**Make observations and gather data**

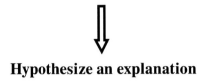

**Hypothesize an explanation**

**Test hypothesis**

**Figure 1.1. The scientific method simplified. This process cycles; it is reevaluated on an ongoing basis.**

- **Inductive reasoning**—inference of general principles based on specific cases; from particular observations to general principles (discovering the rules)—In other words, no matter how much evidence exists for a conclusion, the conclusion could still conceivably be false. Such that,
  *Suppose someone eats 4 oranges out of a box of 100 and finds each of the 4 oranges to be tasty. From this, the person concludes that all the oranges are tasty.*
  The field of geology has many principles based upon inductive reasoning.
  - **Deductive reasoning**—analysis of a specific case based on premises; from general laws to specific predictions (making predictions based on the rules). In other words, if the premises are true, then the conclusion has to be true. Such that,
  *If all Maytag washing machines built in 1969 have switches installed that shutdown the machines any time the lid is opened, and Mabel's Maytag washer was built in 1969, then Mabel's washing machine has a shutdown switch when the lid is opened (or, at least, originally had a shutdown switch if it wasn't removed or jerry-rigged).*

The Power of Science is its Power of Prediction.

# Scientific Principles and Geology

As with the other sciences, the science of geology is based on scientific principles. In the following, the geological principles of parsimony, superposition, and uniformitarianism are defined.

**Parsimony**—basically, in arriving at a hypothesis or course of action, this principle states that "less is better" (i.e., the least complex explanation for an observation is better). The meaning of this approach is best summed up by the acronym KISS ("Keep It Simple, Stupid"). The KISS principle states that simplicity is the key to understanding complex issues.

**Superposition**—this principle (which is basic to the understanding of geologic history) states that in undisturbed sedimentary layers, rocks are deposited in a time sequence, with the oldest on the bottom and youngest on top. Note that in areas where the rocks have been greatly disturbed, it is necessary to determine the tops and bottoms of beds before the normal sequence can be established.

**Uniformitarianism**—is one of the most basic principles of modern geology, the observation that fundamentally the same geological processes that operated today also operated in the distant past. Or, more simply stated, *the present is the key to the past.*

# Geology: Major Divisions

Because the scope of geology is so broad, it has been divided into two major divisions: *physical geology* and *historical geology*. Each of these divisions has been subdivided into a number of more specialized branches.

## PHYSICAL GEOLOGY

Physical geology deals with the study of Earth materials as well Earth's structure, composition, the movements within and upon Earth's surface, and the geologic processes by which Earth's surface is (or has been) changed.

In order to perform detailed studies and gain knowledge in all phases of geology (earth science), the broad division of physical geology includes such basic subbranches as **geochemistry**, the use of chemical principles to study and understand the composition of rocks and environmental quality. The study of the physical properties to infer internal structures is called **geophysics.** The subbranches of **mineralogy**, the study and identification of minerals, and **petrology**, the study of rocks, provide much-needed information about the composition of Earth. In addition, there is **sedimentology** to study sedimentary deposits. Another important branch of physical geology is **structural geology**, the study of the deformation of rocks. **Economic geology**, the exploration and exploitation of geological resources, is also an important branch.

## HISTORICAL GEOLOGY

Historical geology is the study of the origin and evolution of Earth, its environments, and its life (inhabitants) through time. It examines geologic history as a pattern and as process. Like physical geology, historical geology covers such a variety of fields that it has been subdivided into several branches. For example, the geologist utilizes **stratigra-**

phy, which is concerned with the origin, composition, proper sequence, and correlation of the rock strata. **Paleontology**, the study of ancient organisms as revealed by their fossils, provides a background of the development of life on earth, and **paleogeography** affords a mean of studying geographic conditions of past times. Each of these branches (and others)—actually each is a science in itself—make it possible to reconstruct the relations of ancient lands and seas and the organisms that inhabited them.

## Did You Know?

Because physical and historical geologists study the same rocks, the unification of these two important divisions leads ultimately to a better understanding of the composition and history of the Earth.

# Earth's Geological Processes

Earlier it was pointed out that geology is about processes. We can also say that geology is about materials—materials that make up the earth. The materials that make up the earth, of course, are mainly rocks (including dust, silt, sand, soil). Rocks in turn are composed of minerals, and minerals are composed of atoms.

Earth materials and patterns are important and are discussed in detail later in the book but for now it is earth processes that are of interest to us. Earth processes are constantly acting upon and within the earth to change it. Examples of these ongoing processes include formation of rocks, chemical cementation of sand grains together to form rock, construction of mountain ranges, and erosion of mountain ranges. These are internal processes that get their energy from the interior of the earth—most from radioactive decay (nuclear energy). Other examples of ongoing earth processes include those that are more apparent to us (external processes) because they occur relatively quickly and are visible. These include volcanic eruptions, dust storms, mudflows, and beach erosion. The energy source for these processes are solar and gravitational.

It is important to point out that many of these processes are cyclical in nature. The two most important cyclical processes are the hydrologic (water) cycle and the rock cycle.

### HYDROLOGICAL CYCLE

Simply, the water cycle describes how water moves through the environment and identifies the links between groundwater, surface water, and the atmosphere (see figure 1.2). As illustrated, water is taken from the earth's surface into the atmosphere by evaporation from the surface of lakes, rivers, streams, and oceans. This evaporation process occurs when the sun heats water. The sun's heat energizes surface molecules,

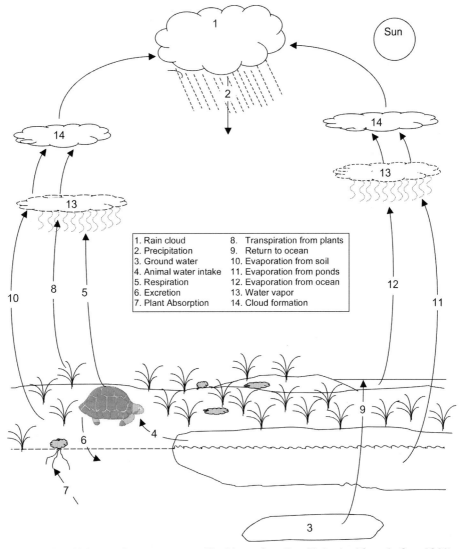

**Figure 1.2.   Water cycle (source: modified from Carolina Biological Supply Co., 1966).**

allowing them to break free of the attractive force binding them together and then evaporate and rise as invisible vapor in the atmosphere.

Water vapor is also emitted from plant leaves by a process called *transpiration*. Every day, an actively growing plant transpires five to ten times as much water as it can hold at once. As water vapor rises, it cools and eventually condenses, usually on tiny particles of dust in the air. When it condenses, it becomes a liquid again or turns directly into a solid (ice, hail, or snow). These water particles then collect and form clouds. The atmospheric water formed in clouds eventually falls to earth as precipitation. The precipitation can contain contaminants from air pollution. The precipitation may fall directly onto surface waters, be intercepted by plants or structures, or fall onto the ground. Most precipitation falls in coastal areas or in high elevations. Some of the water that falls in high elevations becomes runoff water. This water runs over the ground picking up sand, silt, and clay from the soil and carries particles to lower elevations to form streams, lakes, and alluvial fertile valleys.

The water we see is known as *surface water*. Surface water can be broken down into five categories: oceans, lakes, rivers and streams, estuaries, and wetlands.

The health of rivers and streams is directly linked to the integrity of the habitat (and geology) along the river corridor and in adjacent wetlands. Stream quality will deteriorate if activities damage vegetation along riverbanks and in nearby wetlands. Trees, shrubs, and grasses filter pollutants from runoff and reduce soil erosion. Vegetation also provides shade that moderates stream temperature. Stream temperature, in turn, affects the availability of dissolved oxygen in the water for fish and other aquatic organisms.

## ROCK CYCLE

With time and changing conditions, the igneous, sedimentary, and metamorphic rocks of Earth are subject to alteration by the processes of weathering (erosion), volcanism, and tectonism. Known as the **rock cycle** (see figure 1.3), this series of events (or group of changes) represents the response of earth materials to various forms of energy. As shown in figure 1.3, most surface rocks started out as igneous rocks (rocks produced by crystallization from liquid form). When igneous rocks are exposed at the surface they are subject to weathering. Erosion moves particles into rivers and oceans where they are deposited to become sedimentary rocks. Sedimentary rocks can be buried or pushed to deeper levels in the earth, when changes in pressure and temperature cause them to become metamorphic rocks. At high temperatures metamorphic rocks may melt to become magmas. Magmas rise to the surface, crystallize to become igneous rocks, and the processes starts over. Keep in mind that the cycles are not always completed, for there can be many short circuits along the way, as indicated in figure 1.3.

# Did You Know?

All rock that is on earth (except for meteorites) today is made of the same stuff as the rocks that dinosaurs and other ancient life forms walked, crawled or swam over. While the stuff that rocks are made from stays the same, the rocks do not. Over millions of years, rocks are recycled into other rocks. Moving tectonic plates help to destroy and form many types of rocks.

—Jennifer Bergman, 2005

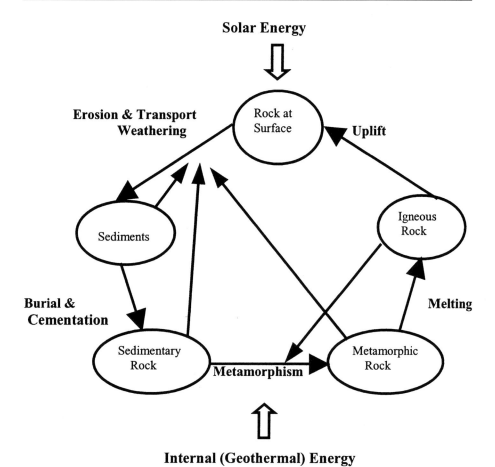

**Figure 1.3.** The rock cycle,—if uninterrupted, will continue completely around the outer margin of the diagram. However, as shown by the arrows, the cycle may be interrupted or "short-circuited" at various points in its course.

## Planet Earth

Earth has a radius of about 6,370 kilometers, although it is about 22 kilometers larger at the equator than at the poles. The circumference of the earth is about 24,874 miles, and the surface area comprises roughly 197 million square miles (about 29 percent) are surface lands. The remaining 71 percent of Earth's surface is covered by water.

Earth, like the other planets in our solar system, revolves around the sun within its own orbit and period of revolution. The earth also rotates on its own axis. The earth rotates from west to east and makes one complete rotation each day. It is this rotating motion that gives us the alternating periods of daylight and darkness that we know as day and night.

The earth also precesses (or wobbles) as it rotates on its axis, much as a top wobbles as it spins. This single wobble has to do with the fact that the earth's axis is tilted at an angle of 23.5 degrees. The tilting of the axis is also responsible for the seasons. Over the years, various hypotheses have been put forward in attempts to determine what the force, or excitation mechanism, is that propels the wobble—such as atmospheric phenomena; continental water storage (changes in snow cover, river runoff, lake levels, or reservoir capacity); and interaction at the boundary of the earth's core and its surrounding mantle, and earthquakes. According to an explanation by NASA (2000), the principal cause of the wobble is fluctuating pressure on the bottom of the oceans, caused by temperature and salinity changes and wind-driven changes in the circulation of the oceans.

Earth revolves around the sun in a slightly elliptical orbit approximately once every 365¼ days. During this solar year, the planet travels at a speed of more than 60,000 miles per hour, and on the average it remains about 93 million miles from the sun.

# Principal Divisions of the Earth

The earth consists of four interconnected "geospheres;" the **atmosphere**, a gaseous envelope surrounding the earth; the **hydrosphere**, the waters filling the depressions and covering almost 71 percent of the earth; the **lithosphere**, the solid part of the earth that underlies the atmosphere and hydrosphere; and the **biosphere**, composed of all living organisms.

## THE ATMOSPHERE

The atmosphere is a thin veil of air, a mixture of gases, that surrounds earth like the skin of an apple—very thin—but very, very vital. The approximate composition of dry air is, by volume at sea level, nitrogen 78 percent, oxygen 21 percent (necessary for life as we know it) argon 0.93 percent, and carbon dioxide 0.03 percent, together with very small amounts of numerous other constituents (see table 1.1). The water vapor content is highly variable and depends on atmospheric conditions. Air is said to be pure when none of the minor constituents is present in sufficient concentration to be injurious to the health of human beings or animals, to damage vegetation, or to cause loss of amenity (e.g., through the presence of dirt, dust, or odors or by diminution of sunshine).

Where does air come from? Genesis 1:2 states that God separated the water environment into the atmosphere and surface waters on the second day of creation. Many scientists state that 4.6 billion years ago a cloud of dust and gases forged the earth and also created a dense molten core enveloped in cosmic gases. This was the **proto-atmosphere** or **proto-air**, composed mainly of carbon dioxide, hydrogen, ammonia, and carbon monoxide, but it did not last long before it was stripped away by a tremendous outburst of charged particles from the sun. As the outer crust of the earth began

**Table 1.1.   Composition of Air/Earth's Atmosphere**

| Gas | Chemical Symbol | Volume (%) |
| --- | --- | --- |
| nitrogen | $N_2$ | 78.08 |
| oxygen | $O_2$ | 20.94 |
| carbon dioxide | $CO_2$ | 0.03 |
| argon | Ar | 0.093 |
| neon | Ne | 0.0018 |
| helium | He | 0.0005 |
| krypton | Kr | trace |
| xenon | Xe | trace |
| ozone | $O_3$ | 0.00006 |
| hydrogen | $H_2$ | 0.00005 |

to solidify, a new atmosphere began to form from the gases outpouring from gigantic hot springs and volcanoes. This created an atmosphere of air composed of carbon dioxide, nitrogen oxides, hydrogen, sulfur dioxide, and water vapor. As the earth cooled, water vapor condensed into highly acidic rainfall, which collected to form oceans and lakes.

For much of Earth's early existence (the first half), only trace amounts of free oxygen were present. But then green plants evolved in the oceans, and they began to add oxygen to the atmosphere as a waste gas. Later, oxygen increased to about 1 percent of the atmosphere and, with time, to its present 21 percent.

How do we know for sure about the evolution of air on Earth? Are we guessing, using "voodoo" science? There is no guessing or voodoo involved with the historical geological record. Consider, for example, geological formations that are dated to 2 billion years ago. In these early sediments there is a clear and extensive band of red sediment ("red bed" sediments)—sands colored with oxidized (ferric) iron. Previous ferrous formations show no oxidation. But there is more evidence. We can look at the time frame of 4.5 billion years ago, when carbon dioxide in the atmosphere was beginning to be lost in sediments. The vast amount of carbon deposited in limestone, oil, and coal indicate that carbon dioxide concentrations must once have been many times greater than the 0.03 percent of today. The first carbonated deposits appeared about 1.7 billion years ago, the first sulfate deposits about 1 billion years ago. The decreasing carbon dioxide was balanced by an increase in the nitrogen content of the air. The forms of **respiration** advanced from fermentation 4 billion years ago to anaerobic **photosynthesis** 3 billion years ago to aerobic photosynthesis 1.5 billion years ago. The aerobic respiration that is so familiar today only began to appear about 500 million years ago. The atmosphere itself continues to evolve, but human activities—with their highly polluting effects—have now overtaken nature in determining the changes. And when you get right down to it, that is one of the overriding themes of this text—human beings and their affect on planet Earth.

The atmosphere is an important geologic agent and is responsible for the processes of weathering that are continually at work on the earth's surface.

## THE HYDROSPHERE

The hydrosphere includes all the waters of the oceans, lakes, and rivers, as well as groundwater—which exists within the lithosphere. Approximately 40 million cubic miles of water cover or reside within the earth. The oceans contain about 97 percent of all water on earth. The other 3 percent is freshwater: (1) snow and ice on the surface of the earth contains about 2.25 percent of the water; (2) usable groundwater is approximately 0.3 percent; and (3) surface freshwater is less than 0.5 percent.

In the United States, for example, average rainfall is approximately 2.6 feet (a volume of 5,900 cubic kilometers). Of this amount, approximately 71 percent evaporates (about 4,200 cubic cm), and 29 percent goes to stream flow (about 1,700 cubic kilometers).

Beneficial freshwater uses include manufacturing, food production, domestic and public needs, recreation, hydroelectric power production, and flood control. Stream flow withdrawn annually is about 7.5 percent (440 cubic kilometers). Irrigation and industry use almost half of this amount (3.4 percent or 200 cubic kilometers per year). Municipalities use only about 0.6 percent (35 cubic kilometers per year) of this amount.

Historically, in the United States, water usage has increased (as might be expected). For example, in 1975, 40 billion gallons of freshwater were used. In 1990, the total increased to 455 billion gallons. Projected use in 2002 is about 725 billion gallons.

The primary sources of freshwater include the following:

1. Captured and stored rainfall in cisterns and water jars
2. Groundwater from springs, artesian wells, and drilled or dug wells
3. Surface water from lakes, rivers, and streams
4. Desalinized seawater or brackish groundwater
5. Reclaimed wastewater

In the United States, current federal drinking water regulations actually define three distinct and separate sources of freshwater. They are surface water, groundwater, and groundwater under the direct influence of surface water (GUDISW). This last classification is the result of the Surface Water Treatment Rule (SWTR). The definition of the conditions that constitute GUDISW, while specific, is not obvious. This classification is discussed in detail later.

## THE LITHOSPHERE

The lithosphere is of prime importance to the geologist. This, the solid, inorganic, rocky crust portion of the earth, is composed of rocks and minerals that, in turn, comprise the continental masses and ocean basins. The rocks of the lithosphere are of three basic types, igneous, sedimentary, and metamorphic.

### *Soil Geology*

We use soil for our daily needs, but we do not sufficiently take account of its slow formation and fast loss. Simply, we take soil for granted. It's always been there—with the implied corollary that it will always be there—right? But where does soil come from?

Of course, soil was formed, and in a never-ending process, it is still being formed. However, as mentioned, soil formation is a slow process—one at work over the course of millennia, as mountains are worn away to dust through bare rock succession.

Any activity, human or natural, that exposes rock to air begins the process. Through the agents of physical and chemical weathering, through extremes of heat and cold, through storms and earthquakes and entropy, bare rock is gradually worn away. As its exterior structures are exposed and weakened, plant life appears to speed the process along.

Lichens cover the bare rock first, growing on the rock's surface, etching it with mild acids and collecting a thin film of soil that is trapped against the rock. This process changes the conditions so much that the lichens can no longer survive on the rock and are replaced by mosses.

The mosses establish themselves in the soil trapped and enriched by the lichens and collect even more soil. They hold moisture to the surface of the rock, setting up another change in environmental conditions.

Well-established mosses hold enough soil to allow herbaceous plant seeds to invade the rock. Grasses and small flowering plants move in, sending out fine root systems that hold more soil and moisture and work their way into minute fissures in the rock's surface. More and more organisms join the increasingly complex community.

Weedy shrubs are the next invaders, with heavier root systems that find their way into every crevice. Each stage of succession affects the decay of the rock's surface and adds its own organic material to the mix. Over the course of time, mountains are worn away, eaten away to soil, as time, plants, and weather work on them.

The parent material, the rock, becomes smaller and weaker as the years, decades, centuries and millennia go by, creating the rich, varied, and valuable mineral resource we call soil.

Perhaps no term causes more confusion in communication between various groups of average persons, soil geologists, soil scientists, soil engineers, and earth scientists than the word *soil*. In simple terms, soil can be defined as the topmost layer of decomposed rock and organic matter, which usually contains air, moisture, and nutrients and can therefore support life. Most people would have little difficulty in understanding and accepting this simple definition. Then why are various groups confused on the exact meaning of the word *soil*? Quite simply, confusion reigns because soil is not simple—it is quite complex. In addition, the term *soil* has different meanings to different groups (like pollution, the exact definition of soil is a personal judgment call). Let's take a look at how some of these different groups view soil.

*The average person* seldom gives soil a first or second thought. Why should he? Soil isn't that big a deal—that important—it doesn't impact his life, pay his bills, or feed his bulldog, right?

Not exactly. Not directly.

The average person seldom thinks about soil as soil. He or she may think of soil in terms of dirt, but hardly ever as soil. Why is this? Having said the obvious about the confusion between soil and dirt, let's clear up this confusion.

First of all, soil is not dirt. Dirt is misplaced soil—soil where we don't want it—contaminating our hands and fingernails, messing up our clothes and automobiles, and tracking up our floors. Dirt we try to clean up and keep out our living environments.

Secondly, soil is too special to be called dirt. Why? Because soil is mysterious and, whether we realize it or not, essential to our existence. Because we think of it as common, we relegate soil to an ignoble position. As our usual course of action, we degrade it, abuse it, throw it away, contaminate it, ignore it—we treat it like dirt. Only feces holds a more lowly status than it does. Soil deserves better.

Why?

Again, because soil is not dirt—how can it be? It is not filth, or grime, or squalor. Instead soil is clay; air; water; sand; loam; the organic detritus of former life-forms (including humans); and, most important, the amended fabric of Earth itself; if water is Earth's blood and air is Earth's breath, then soil is its flesh and bone and marrow. Simply put, soil is the substance that most life depends on.

*Soil scientists* (or pedologists) are groups interested in soils as a medium for plant growth. Their focus is on the upper meter or so beneath the land surface (this is known as the weathering zone, which contains the organic-rich material that supports plant growth) directly above the unconsolidated parent material. Soil scientists have developed a classification system for soils based on the physical, chemical, and biological properties that can be observed and measured in the soil.

*Soil engineers* are typically soil specialists who look at soil as a medium that can be excavated using tools. Soil engineers are not concerned with the plant-growing potential of a particular soil but rather are concerned with a particular soil's ability to support a load. They attempt to determine (through examination and testing) a soil's particle size, particle-size distribution, and the plasticity of the soil.

*Earth scientists* (or geologists) have a view that typically falls between pedologists and soil engineers—they are interested in soils and the weathering processes as past indicators of climatic conditions and in relation to the geologic formation of useful materials ranging from clay deposits to metallic ores.

Would you like to gain new understanding of soil? Take yourself out to a plowed farm field somewhere, anywhere. Reach down and pick up a handful of soil, and look at it—really look at it closely. What are you holding in your hand? Look at the two descriptions that follow and you may gain a better understanding of what soil actually is and why it is critically important to us all.

1. A handful of soil is alive, a delicate living organism as lively as an army of migrating caribou and as fascinating as a flock of egrets. Literally teeming with life of multitudinous forms, soil deserves to be classified as an independent ecosystem, or more correctly stated, as many ecosystems.
2. When we pick up a handful of soil, exposing Earth's stark bedrock surface, it

should remind us (and maybe startle us to the realization) that without its thin, living soil layer, Earth is a planet as lifeless as our own moon (Spellman, 1998).

## THE BIOSPHERE

Thanks to the life-giving qualities of air and water, earth is populated by countless species of plants and animals. This horde of organisms comprises the biosphere. Most of the planet's life is found from three meters below the ground to thirty meters above it and in the top 200 meters of the oceans and seas.

In regards to the life-forms that make up the biosphere, have you ever asked what life is? What does it mean to be alive? Have you ever tried to define life? If so, how did you define it? If these questions strike you as odd, consider them for a moment (they are almost as difficult as defining the origin of life). Of course we all have an intuitive sense of what life is, but if you had difficulty, as is probably the case, with answering these questions, you are not alone. These questions are open to debate and have been from the beginning of time. One thing is certain; life is not a simple concept, and it is impossible to define.

Along with the impossibility of defining life definitively, it is not always an easy thing to tell the difference between living, dead, and nonliving things. Prior to the seventeenth century, many people believed that nonliving things could spontaneously turn into living things. For example, it was believed that piles of straw could turn into mice. Obviously, that is not the case. There are some very general rules to follow when trying to decide if something is living, dead, or nonliving. Scientists have identified seven basic characteristics of life. Keep in mind that for something to be described as living, it must display *all seven* of these characteristics (i.e., "characteristic" is plural). Although many of us have many different opinions about what "living" means, the following characteristics were designated "characteristics of living things" with the consensus of the scientific community.

- **Living things are composed of cells**: Living things exhibit a high level of organization, with multicellular organisms being subdivided into cells, and cells into organelles, and organelles into molecules, and so on.
- **Living things reproduce**: All living organisms reproduce, either by sexual or asexual means.
- **Living things respond to stimuli**: All living things respond to stimuli in their environment.
- **Living things maintain homeostasis**: All living things maintain a state of internal balance in terms of temperature, pH, water concentrations, and so on.
- **Living things require energy**: Some view life as a struggle to acquire energy (from sunlight, inorganic chemicals, or another organism), and release it in the process of forming adenosine triphosphate (ATP). The conventional view is that living organisms require energy, usually in the form ATP. They use this energy to carry out energy-requiring activities such as metabolism and locomotion.

- **Living things display heredity**: Living organisms inherit traits from the parent organisms that created them.
- **Living things evolve and adapt**: All organisms have the ability to adapt or adjust to their surroundings. An example of this might be adapting to environmental change resulting in an increased ability to reproduce.

*Interesting Point*: Again, if something follows one or just a few of the characteristics listed above, it does not necessarily mean that it is living. To be considered alive, an object must exhibit *all* of the characteristics of living things. A good example of a nonliving object that displays at least one characteristic for living is sugar crystals growing on the bottom of a syrup dispenser. On the other hand, there is a stark exception to the characteristics above in the example of the mule: mules cannot reproduce because they are sterile. Another nonliving object that exhibits many of the characteristics of life is a flame. Think about it; a flame:

- respires
- requires nutrition
- reproduces
- excretes
- grows
- moves
- is irritable
- is organized

We all know that a flame is not alive, but how do we prove that to the skeptic? The best argument we can make is

1. Nonliving materials never replicate using DNA and RNA (hereditable materials).
2. Nonliving material cannot carry out anabolic metabolism.

# Key Terms

After reviewing the characteristics of life listed above, it should be obvious that we need to follow Voltaire's advice; that is, "If you wish to converse with me, please define your terms." Therefore, many of the terms used to describe the characteristics of life or related to the characteristics are defined here.

**Aerobic**—occurring or living only in the presence of oxygen.
**Anabolism**—the utilization of energy and materials to build and maintain complex structures from simple components.
**Anaerobic**—active, living, occurring, or existing in the absence of free oxygen.
**Amino acids**—building blocks of proteins.
**Asexual reproduction**—requires one parent cell.

**Autotrophic**—using light energy (photosynthesis) or chemical energy (chemosynthesis).

**Catabolism**—the breaking down of complex materials into simpler ones using enzymes and releasing energy.

**Carbohydrates**—main source of energy for living things such as sugar and starch.

**Cellular basis of life**—cells are the basis of life. There are two types of cells: prokaryotes and eukaryotes.

**Cold-blooded animals**—body temperatures change with the environment.

**Compound**—two or more elements chemically combined.

**Digestion**—process by which food is broken down into simpler substances.

**DNA**—the double helix of DNA is the unifying chemical of life; its linear sequence defines the diversity of living things.

**Elements**—pure substance that cannot be broken down into simpler substances.

**Enzymes**—special types of proteins that regulate chemical activities.

**Excretion**—process of getting rid of waste materials.

**Eukaryotes**—these are cells with a nucleus; they are found in human and other multicellular organisms (plants and animals), also algae, protozoa.

**Evolution**—the modification of species, the core theme of biology.

**Food**—needed by living things to grow, develop, and repair body parts.

**Heterotrophic**—obtaining materials and energy by the breaking down of other biological material using digestive enzymes and then assimilating the usable byproducts.

**Ingestion**—taking in food or producing food.

**Inorganic compounds**—may or may not contain carbon.

**Life span**—maximum length of time an organism can be expected to live.

**Lipids**—energy-rich compounds made of carbon, oxygen, and hydrogen.

**Metabolism**—chemical reactions that occur in living things.

**Movement**—nonliving material moves only as a result of external forces, while living material moves as a result of internal processes at cellular level or at organism level (locomotion in animals and growth in plants).

**Nucleic acids**—store information that helps the body make proteins it needs.

**Organic compounds**—found in living things and contain carbon.

**Organism**—any living thing.

**Prokaryotes**—cells without a nucleus, including bacteria and cyanophytes (blue-green algae). The genetic material is a single circular DNA and is contained in the cytoplasm, since there is no nucleus.

**Proteins**—used to build and repair cells; made of amino acids.

**Respiration**—taking in oxygen and using it to produce energy.

**Response**—action, movement, or change in behavior caused by a stimulus.

**Sexual reproduction**—requires two parent cells.

**Stimulus**—signal that causes an organism to react.

**Structure and function**—at all levels of organization, biological structures are shaped by natural selection to maximize their ability to perform their functions.

**Unity in diversity**—explained by evolution, all organisms linked to their ancestors through evolution; scientists classify life on Earth into groups related by ancestry;

related organisms descended from a common ancestor and have certain similar characteristics.

**Warm-blooded animals**—maintain a constant body temperature.

## Did You Know?

All four divisions of the Earth can be and often are present in a single location. For example, a piece of soil will of course have mineral material from the lithosphere. Additionally, there will be elements of the hydrosphere present as moisture within the soil, the biosphere as insects and plants, and even the atmosphere as pockets of air between soil pieces.

## Structure of Earth

Earth is made up of three main compositional layers: crust, mantle, and core (see figure 1.4). The crust has variable thickness and composition: Continental crust is 10–50 kilometers thick while the oceanic crust is 8–10 kilometers thick. The elements silicon, oxygen, aluminum, and iron make up the earth's crust. Like the shell of an egg, the earth's crust is brittle and can break. Earth's continental crust is 35 kilometers thick. The oceanic crust is 7 kilometers thick.

Based on seismic (earthquake) waves that pass through the earth, we know that below the crust is the mantle, a dense, hot layer of semisolid (plastic-like liquid) rock approximately 2,900 kilometers thick. The mantle, which contains silicon; oxygen; aluminum; and more iron, magnesium, and calcium than the crust, is hotter and denser because temperature and pressure inside the earth increase with depth. According to USGS (1999), as a comparison, the mantle might be thought of as the white of a boiled egg. The 30-kilometer-thick transitional layer between the mantle and crust is called the Moho layer. The temperature at the top of the mantle is 870°C. The temperature at the bottom of the mantle is 2,200°C.

At the center of the earth lies the core, which is nearly twice as dense as the mantle because its composition is metallic [Iron (Fe)-Nickel (NI) alloy] rather than stony. Unlike the yolk of an egg however, the earth's core is actually made up of two distinct parts: a 2,200-kilometer-thick liquid outer core and a 1,250-kilometer-thick solid inner core. As the earth rotates, the liquid outer core spins, creating the earth's magnetic field.

## Chapter Review Questions

1.1   How thick is earth's continental crust? _____ km

1.2   How thick is earth's oceanic crust? _____ km

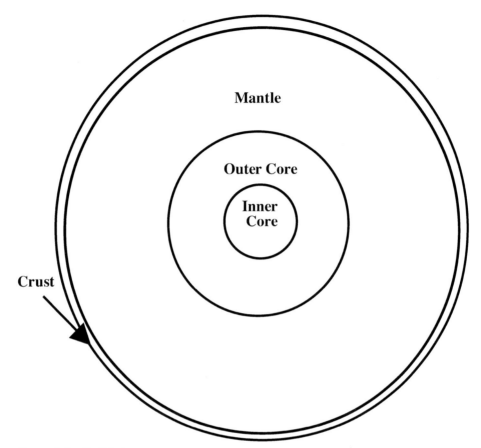

**Figure 1.4.  Earth's layers.**

1.3   The layer between the mantle and the crust is known as _____.
1.4   Plasticity describes magma's ability to _____.
1.5   What is the first step in the scientific method?
1.6   In geology, _____ means less is better.
1.7   In geology, _____ means the present is the key to the past.
1.8   _____ describes how water moves through the environment.
1.9   The solid part of the earth is known as the _____.
1.10  Define dirt.

# References and Recommended Reading

*Air Quality Criteria*. 1968. Staff report, Subcommittee on Air and Water Pollution, Committee on Public Works, U.S. Senate, 94–411.

*American Heritage Dictionary of the English Language*, 4th ed. 2000. Boston: Houghton Mifflin.

Bergman, J. 2005. *Rocks and the Rock Cycle*. Accessed 05/23/08 at www.windows.ucar.edu/tour/ling=/earth/geology/rocks_intro.html.

Blue, J. 2007. Descriptor terms *Gazetteer of Planetary Nomenclature*. USGS. Accessed 05/20/08 at http://planetarynames.wr.usgs.gov/jsp/append5.jsp.

Campbell, N. A. 2004. *Biology: Concepts and Connections*, 4th CD-ROM ed. Menlo Park, CA: Benjamin-Cummings.

Goshorn, D. 2006. *Proceedings—DELMARVA Coastal Bays Conference III: Tri-State Approaches to Preserving Aquatic Resources*. Washington, DC: USEPA.

Huxley, T. H. 1876. *Science and Education*, Volume III, *Collected Essays*. New York: D. Appleton & Company.

Jones, A. M. 1997. *Environmental Biology*. New York: Routledge.

Keeton, W. T. 1996. *Biological Science*. Kingston, MA: R. S. Means.

King, R. M. 2003. *Biology Made Simple*. New York: Broadway Books.

Koch, R. 1882. Uber die Atiologie der Tuberkulose. In *Verhandlungen des Knogresses fur Innere Medizin*. Wiesbaden: Erster Kongress.

Larsson, K. A. 1993. Prediction of the pollen season with a cumulated activity method, *Grana*, 32,111–14.

Med Net. 2006. *Definition of Koch's Postulates*. Medicine Net.com.

NASA. 2000. *A Mystery of Earth's Wobble Solved: It's the Ocean*. Accessed 05/24/08 at www.jpl.nasa.gov/releases/2000/chandlerwobble.html.

Press, R., and Siever, F. 2001. *Earth*, 3rd ed. New York: W. H. Freeman & Co.

SGM. 2006. *The Scientific Method, Fish Health and Pfiesteria*. Adelphi: University of Maryland; NOAA.

Spellman, F. R. 1998. *Environmental Science and Technology: Concepts and Applications*, Lancaster, PA: Technomic Publishing Company.

Spellman, F. R. 2008. *The Science of Air: Concepts and Applications*. Boca Raton, FL: CRC Press.

Spellman, F. R., and Whiting, N. E. 2006. *Environmental Science and Technology*, 2nd ed. Rockville, MD: Government Institutes.

Spieksma, F. T. 1991. Aerobiology in the nineties: Aerobiology and pollinosis, *International Aerobiology Newsletter*, 34, 1–5.

USEPA. 2006. What is the scientific method? Accessed at www.epa.gov/maia/html/scientific.html.

# CHAPTER 2

# Minerals

*The Earth is just too small and fragile a basket for the human race to keep all its eggs in.*

—Robert Heinlein

The geologist is primarily interested in the earth's rocky crust. Because rocks are aggregates of minerals, in order to study and understand the rocks that make up earth's crust it is necessary to know something about minerals. Minerals are solid chemical elements or compounds that occur naturally within the crust of the earth. They are inorganic and have a definite chemical composition or range of composition, a specified orderly internal arrangement of atoms (crystalline structure), and certain other distinct physical properties.

Again, the earth is composed of rocks. Rocks are aggregates of minerals. Minerals are composed of atoms. In order to understand rocks, we must first must have an understanding of minerals. In order to understand minerals, we must have some basic understanding of atoms—what they are and how they interact with one another to form minerals.

## Matter

Although a detailed discussion of chemistry is not within the scope of this book,[1] an introduction to chemical terminology and atoms is necessary if we are to understand the chemical composition of minerals.

Just as we use several adjectives to describe an object (color of object, how tall or short, how bulky or thin, how light or heavy, etc.) several characteristics must be used in combination to adequately describe a kind of matter. However, simply saying that something is a colorless solid isn't enough to identify it as diamond. A lot of solids are colorless, for example, some quartz and feldspar, as well as many rock aggregates. More details are needed before we can zero in on the identity of a substance. Chemists will, therefore, determine several **properties**, both chemical and physical, in order to characterize a particular sample of matter—to distinguish one form of matter from another. The following discussion describes the differences between the two kinds of properties, chemical and physical.

## DEFINING MATTER

A thorough understanding of matter—how it consists of elements that are built from atoms—is critical for grasping chemistry. **Matter** is anything that occupies space and has weight (mass). Matter (or mass-energy) is neither created nor destroyed during chemical change.

As mentioned, along with the properties of having weight and taking space, matter has **chemical** and **physical properties**. These properties are actually used to describe **substances**, which are definite varieties of matter. Copper, gold, salt, sugar, and rust are all examples of substances. All of these substances are uniform in their makeup. However, if we pick up a common rock from our garden, we cannot call the rock a substance because it is a mixture of several different substances.

## Key Points!

Matter is measured by making use of its two properties. Anything that has the properties of having weight and taking up space *must* be matter.

A *substance* is a definite variety of matter, all specimens of which have the same properties.

## PHYSICAL PROPERTIES

Substances have two kinds of physical properties: intensive and extensive. **Intensive physical properties** include those features that definitely distinguish one substance from another. Intensive physical properties do not depend on the amount of the substance (matter present). Some of the important intensive physical properties are color, taste, melting point, boiling point, density, luster, and hardness. It is important to note that it takes a combination of several intensive properties to identify a given substance. For example, a certain substance may have a particular color that is common to it but not necessarily unique to it. A white diamond is, as its name implies, white. However, is another gemstone that is white and faceted to look like a diamond really a diamond? Remember that a diamond is one of the hardest known substances. To determine if the white faceted gemstone is really a diamond we would also have to test its hardness and not rely on its appearance alone.

Some of the important intensive physical properties are defined below.

**Density.** Density is the mass per unit volume of a substance (mass of a substance divided by its volume). Suppose we had a cube of lard and a large box of crackers, each having a mass of 400 grams. The density of the crackers would be much less than the density of the lard because the crackers occupy a much larger volume than the lard occupies. The density of an object or substance can be calculated by using the formula:

$$\text{Density} = \frac{\text{mass}}{\text{volume}}$$

Starting with water as one example, consider that perhaps the most common measures of density are pounds per cubic foot (lb/ft³) and pounds per gallon (lb/gal). However, the density of a dry substance, such as sand, lime, and soda ash, is usually expressed in pounds per cubic foot. The density of a gas, such as methane or carbon dioxide, is usually expressed in pounds per cubic foot. As shown in table 2.1, the density of a substance like water changes slightly as the temperature of the substance changes. This happens because substances usually increase in volume (size), as they become warmer. Because of this expansion with warming, the same weight is spread over a larger volume, so the density is lower when a substance is warm than when it is cold.

**Specific Gravity.** Also known as SG, specific gravity is a unitless measure of the weight of a substance compared to the weight of an equal volume of water—a substance having a specific gravity of 2.5 weighs two and a half times more than water. This relationship is easily seen when a cubic foot of water, which weighs 62.4 pounds, is compared to a cubic foot of aluminum, which weighs 178 pounds. Aluminum is 2.7 times as heavy as water.

Gold can easily be distinguished from "fool's gold" by specific gravity alone because it is not that difficult to find the specific gravity of a piece of metal. All we need do is to weigh the metal in air, then weigh it under water. Its loss of weight is the weight of an equal volume of water. To find the specific gravity, divide the weight of the metal by its loss of weight in water.

**Table 2.1.   Water Weight and Density Relative to Temperature**

| Temperature (°F) | Specific Weight (lb/ft³) | Density (slugs/ft³) |
|---|---|---|
| 32 | 62.4 | 1.94 |
| 40 | 62.4 | 1.94 |
| 50 | 62.4 | 1.94 |
| 60 | 62.4 | 1.94 |
| 70 | 62.3 | 1.94 |
| 80 | 62.2 | 1.93 |
| 90 | 62.1 | 1.93 |
| 100 | 62.0 | 1.93 |
| 110 | 61.9 | 1.92 |
| 120 | 61.7 | 1.92 |
| 130 | 61.5 | 1.91 |
| 140 | 61.4 | 1.91 |
| 150 | 61.2 | 1.90 |
| 160 | 61.0 | 1.90 |
| 170 | 60.8 | 1.89 |
| 180 | 60.6 | 1.88 |
| 190 | 60.4 | 1.88 |
| 200 | 60.1 | 1.87 |
| 210 | 59.8 | 1.86 |

$$\text{Specific gravity} = \frac{\text{Weight of a substance}}{\text{Weight of equal volume of water}}$$

Here's an example of finding specific gravity. Problem: Suppose a piece of tin weighs 110 pounds in air and 74 pounds under water. What is the specific gravity?
    Solution:

(1)  Step 1: 110 lbs subtract 74 lbs = 36 lbs loss of weight in water
(2)  Step 2:

$$\text{Specific gravity} = \frac{110 \text{lbs}}{36 \text{ lbs}} = 3.1$$

# Key Point!

In a calculation of specific gravity, it is *essential* that the densities be expressed in the same units.

The specific gravity of water is one, which is the standard, the reference for which all other substances (i.e., liquids and solids) are compared. Any object that has a specific gravity greater than one will sink in water. Considering the total weight and volume of a ship, its specific gravity is less than one; therefore, it can float.

In geology, specific gravity has application. For example, it has application in regards to the earth's crust, which is composed mostly of the minerals calcite, feldspar, and quartz. These minerals have SGs around 2.75, and this is close to the average SG of the rocks on the outer surface of the earth's crust. Thus, the SG of most rocks that are initially picked up and later sorted out is familiar to the geologist; they have a specific gravity of approximately 2.75.

**Hardness.** Hardness is commonly defined as a substance's relative ability to resist scratching or indentation. Hardness is one of the easiest ways to distinguish one mineral from another. Actual hardness testing involves measuring how far an "indenter" can be pressed into a given material under a known force. A substance will scratch or indent any other substance that is softer. Table 2.2 is used for comparing the hardness of mineral substances.

Do not confuse mineral hardness for water hardness. In water treatment, *hardness* is a characteristic of water caused primarily by calcium and magnesium ions. Water hardness can cause many maintenance problems, especially with piping and process components where scale buildup can occur.

**Streak.** Streak is the color of a crushed mineral's powder that is left when the mineral is rubbed across a piece of unglazed tile, leaving a line similar to a pencil or crayon mark. This line is composed of the powdered minerals. The color of this powdered material is known as the streak of the mineral, and the unglazed tile used

**Table 2.2.   Moh's Hardness Scale**

| Hardness (H) | Mineral |
| --- | --- |
| 1 | Talc |
| 2 | Gypsum (fingernail: H = 2.5) |
| 3 | Calcite (penny: H = 3) |
| 4 | Fluorite |
| 5 | Apatite |
| 6 | Feldspar (glass plate: H = 5.5) |
| 7 | Quartz |
| 8 | Topaz |
| 9 | Corundum |
| 10 | Diamond (hardest) |

in such a test called a streak plate. Most light-colored, nonmetallic minerals have a white or colorless streak, as do most silicates, carbonates, and most transparent minerals. The streak test is most useful for identifying dark-colored minerals, especially metals.

**Color.** Unless we are color-blind, probably one of the first things we notice about a mineral is its color. Most of us are familiar with the colors of various substances. Pure water is usually described as colorless. Water takes on color when foreign substances such as organic matter from soils, vegetation, minerals, and aquatic organisms are present. Like water, the same mineral may vary greatly in color from one specimen to another, and with certain exceptions, color is of limited use in mineral identification. When using color in mineral identification, we must consider whether the specimen is being examined in natural or artificial light; whether the surface being examined is weathered or fresh; and whether the mineral is wet or dry.

**Luster.** The appearance of the surface of a mineral as seen in reflected light is called luster. Terms such as shining (bright), glistening (spackling brightness), splendent (glossy brilliance), and dull (lacking luster) are commonly used to indicate the degree of luster present. Some minerals shine like the metals gold and silver. These are said to have metallic luster. A mineral does not have to be a metal to have luster. Other lusters are called nonmetallic. Some common examples are shown below:

• Adamantine—brilliant glossy luster (diamonds)
• Vitreous—glassy, looks like glass (topaz or quartz)
• Resinous—the luster of resin (sphalerite)
• Greasy—like an oily surface (nepheline)
• Pearly—like mother-of-pearl (talc)
• Silky—luster of silk or rayon (asbestos)
• Dull—as the name implies (chalk or clay)

**Malleability.** When a mineral, such as gold or copper, can be hammered into thin sheets it is said to be malleable.

**Ductility.** When a substance (e.g., copper, gold, or silver) can be drawn into thin wires, it is said to be ductile.

**Conductivity.** The ability of a substance (copper, aluminum, gold, silver, etc.) to conduct electricity is called conductivity.

**Boiling point.** The temperature at which the vapor pressure of a liquid is equal to the pressure on the liquid (e.g., atmospheric pressure) is known as the boiling point.

**Melting/freezing point.** The temperature at which the solid and liquid phases of a substance are in equilibrium at atmospheric pressure is known as the melting/freezing point. Extensive physical properties are such features as *mass, weight, volume, length,* and *shape.* Extensive physical properties are dependent upon the amount of the substance (matter present).

## CHEMICAL PROPERTIES

The nonchemist often has difficulty in distinguishing the physical versus the chemical properties of a substance. One test that can help is to ask the question: Are the properties of a substance determined without changing the identity of the substance? If we answer *yes,* then the substance is distinguished by its physical properties. If we answer *no,* then we can assume the substance is defined by its chemical properties. Simply, the **chemical properties** of a substance describe its ability to form new substances under given conditions.

To determine the chemical properties of certain substances we can observe how the substance reacts in the presence of air, acid, water, a base, and other chemicals. We can also observe what happens when the substance is heated. If we observe a change from one substance to another, we know that a **chemical change**, or a **chemical reaction**, has taken place.

# Key Point!

The chemical properties of a substance may be considered to be a listing of all the chemical reactions of a substance and the conditions under which the reactions occur.

Another example can be used to demonstrate the difference between physical and chemical change. When a carpenter cuts pieces of wood from a larger piece of wood to build a wooden cabinet, the wood takes on a new appearance. The value of the crafted wood is increased as a result of its new look. This kind of change, in which the substance remains the same, but only the appearance is different, is called a **physical change**. When this same wood is consumed in a fire, however, ashes result. This change of wood into ashes is called a **chemical change.** *In a chemical change a new substance is produced.* Wood has the property of being able to burn. Ashes cannot burn.

**KINETIC THEORY OF MATTER**

There are three states of matter—*solids, liquids, and gases*. All matter is made of *molecules*. Matter is held together by attractive forces, which prevent substances from coming apart. The molecules of a solid are packed more closely together and have little freedom of motion. In liquids, molecules move with more freedom and are able to flow. The molecules of gases have the greatest degree of freedom and their attractive forces are unable to hold them together.

The **Kinetic Theory of Matter** is a statement of how we believe atoms and molecules behave, and it relates to the ways we look at the things around us. Essentially, the theory states that all molecules are always moving. More specifically, the theory says:

- All matter is made of atoms, the smallest bit of each element. A particle of a gas could be an atom or a group of atoms.
- Atoms have energy of motion that we feel as temperature. At higher temperatures, the molecules move faster.

Matter changes its state from one form to another. Examples of how matter changes its state include

- melting, the change of a solid into its liquid state
- freezing, the change of a liquid into its solid state
- condensation, the change of a gas into its liquid form
- evaporation, the change of a liquid into its gaseous state
- sublimation, the change of a gas into its solid state and vice versa (without becoming liquid)

# Atomic Theory

Up to this point we have described how chemical change involves a complete transformation of one substance into another. Now that we have established a basic understanding of the chemical change of substances, we need to look at the structure of matter.

The ancient Greek philosophers believed that all the matter in the universe was composed of four elements: earth, fire, air, and water. Today we know of more than 100 different elements, and scientists are attempting to create additional elements in their laboratories. An **element** is a substance from which no other material can be obtained.

Today, we know that **atoms** are the basic building blocks of all *matter*. Atoms are the smallest particles of matter with constant properties.

Atoms are so small that it would take approximately 2 billion atoms side by side to equal one meter in length!

Atoms are so small that scientists were forced to devise special weights and measures for them.

- Mass/weight: **atomic units (au)**
  1 au $= 1.6604 \times 10^{-24}$ g
- Length: **Angstrom (Å)**
  1 Å $= 10^{-8}$ cm

It is interesting to note that scientists used to believe that atoms were *indivisible,* but we now know that they are made up of many **subatomic particles.** Chemistry primarily deals with three subatomic particles: **protons, neutrons,** and **electrons.**

- Protons and neutrons are located in the **nucleus** (center) of the atom.
- Electrons are located in **orbitals** around the nucleus (see figure 2.1).

| Particle | Charge | Mass |
|----------|--------|------|
| Proton (P) | + | 1 atomic unit |
| Neutron (N) | no charge | 1 atomic unit |
| Electron (e) | − (none) | |

An atom contains subatomic particles including:

- protons ( + ) positively charged particles
- electrons ( − ) negatively charged particles
- neutrons (0) having no charge

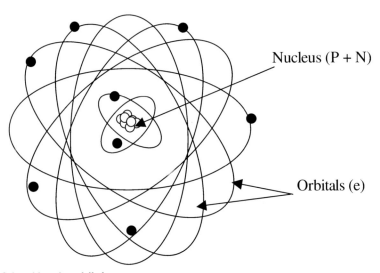

**Figure 2.1.  Atomic orbitals.**

The **stability** of a nucleus depends on the balance between:

- Attractive gravitational forces

- Repulsive electronic forces

- The ratio of protons to neutrons

## Key Point!

In a stable atom (or neutral atom), the number of atoms = the number of protons.

### ELEMENTS

As mentioned, scientists have identified more than 100 different types of atoms, which they call **elements.** An atom is the smallest unit of an element that still retains the properties of that element. An **element** is a substance from which no other material can be obtained. Stated differently, the chemical elements are the simplest substances into which ordinary matter may be divided. All other materials have more complex structures and are formed by the combination of two or more of these elements.

Of the more than 100 elements, only 83 are not naturally radioactive, and of those only 50 or so are common enough to our experience to be useful in this text. These elements, though, are going to stay the same for a long time; that is, they will outlast any political entity.

Elements are substances; therefore they have physical properties that include density, hardness, odor, color, and chemical properties that describe their ability to form new substances (i.e., a list of all the chemical reactions of that material under specific conditions).

Each element is represented by a **chemical symbol.** Chemical symbols are usually derived from the element's name (e.g., Al for aluminum). The chemical symbols for the elements known in antiquity are taken from their Latin names (e.g., Pb for lead). For every element there is one and only one uppercase letter (e.g., O for oxygen). There may or may not be a lowercase letter with it (e.g., Cu for copper). When written in chemical equations, we represent the elements by the symbol alone, with no charge attached.

### Examples of Chemical Symbols

Fe (Iron)            P (Phosphorus)
Al (Aluminum)        Ag (Silver)

Ca (Calcium)              Cl (Chlorine)
C (Carbon)                H (Copper)
N (Nitrogen)              K (Potassium)
Rn (Radon)                He (Helium)
H (Hydrogen)              Si (Silicon)
Cd (Cadmium)              U (Uranium)

Presently, as mentioned, we know only about 100 elements, but well over a million compounds. Only 88 of the known elements are present in detectable amounts on Earth, and many of these 88 are rare. Only 10 elements make up approximately 99 percent (by mass) of the earth's crust, including the surface layer, the atmosphere, and bodies of water (see table 2.3).

As can be seen from table 2.3, the most abundant element on Earth is oxygen, which is found in the free state in the atmosphere as well as in combined form with other elements in numerous minerals and ores.

## MATTER AND ATOMS

**Molecules** consist of two or more **atoms** that have chemically combined

$$A + B \rightarrow C$$

When atoms (elements) chemically combine to form **compounds**, they *lose all their original properties* and create a new set of properties, unique to the compound (see figure 2.2). For example, sodium (Na) and chlorine (Cl) are poisonous, but they combine to form a compound called sodium chloride, which is ordinary table salt.
   *A molecule is the smallest unit of a compound that still retains the properties of that compound* (see figure 2.3).

**Table 2.3.   Elements Making 99 Percent of Earth's Crust, Oceans, and Atmosphere**

| Element | Symbol | % of Composition | Atomic Number |
|---|---|---|---|
| Oxygen | O | 49.5% | 8 |
| Silicon | Si | 25.7% | 14 |
| Aluminum | Al | 7.5% | 13 |
| Iron | Fe | 4.7% | 26 |
| Calcium | Ca | 3.4% | 20 |
| Sodium | Na | 2.6% | 11 |
| Potassium | K | 2.4% | 19 |
| Magnesium | Mg | 1.9% | 12 |
| Hydrogen | H | 1.9% | 1 |
| Titanium | Ti | 0.58% | 22 |

**Element**

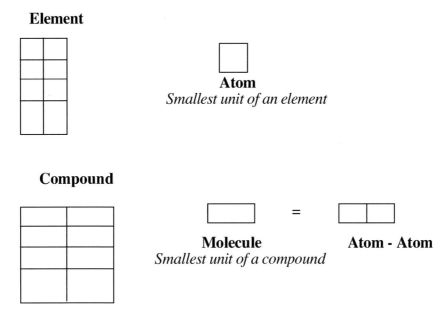

**Atom**
*Smallest unit of an element*

**Compound**

**Molecule**
*Smallest unit of a compound*

**Atom - Atom**

**Figure 2.2.   Element-atom; compound-molecule = atom.**

*One molecule of the compound called water*

**Figure 2.3.   Water molecule.**

Each individual *compound* will always contain *the same elements in the same proportions* by weight.

$$H_2O \neq H_2O_2$$
Water    hydrogen peroxide

Compounds are represented by **chemical formulas**. Chemical formulas consist of *chemical symbols* and *subscripts* that describe the relative number of atoms present in each compound. Thus, a chemical formula tells us how many atoms of each element are in a molecule of any substance. As mentioned, the chemical formula for water is $H_2O$. The "H" is the symbol for hydrogen. Hydrogen is a part of the water molecule. The "O" means that oxygen is also part of the water molecule. The subscript "2" after the H means that two atoms of hydrogen are combined with one atom of oxygen in each water molecule.

**Common Chemical Formulas** include

- $H_2O$ (water)
- NaCl (sodium chloride, table salt)
- HCl (hydrochloric acid)
- $CCl_4$ (carbon tetrachloride)
- $CH_2Cl_2$ (methylene chloride)

Most naturally occurring matter consists of **mixtures** of elements and/or compounds. Mixtures are found in rocks, the ocean, vegetation, just about anything we find. **Mixtures** are combinations of elements and/or compounds held together by physical rather than chemical means. It is important to note that *mixtures are physical combinations* and *compounds are chemical combinations*.

Mixtures have a wide variety of compositions and can be separated into their ingredients by physical means (e.g., filtering, sorting, distillation, etc.). The *components of a mixture retain their own properties*.

The thing to remember about mixtures is that we start with some pieces, combine them, and we can do something to pull those pieces apart again. We wind up with the same molecules (in the same amounts) that we started with. Two classic examples of mixtures are concrete and salt water. When mixed, they seem to form compounds, but because physical forces hold them together, physical forces can take them apart (cement can still have the basic parts removed by grinding and salt water by filtering). The parts are then just like they were when we started.

**Compounds** are two or more elements that are "stuck" (bonded) together in definite proportions by a chemical reaction, for example, water ($H_2O$) and halite (table salt—NaCl). Compounds always have uniform proportions, have unique properties that are different than their components, and *cannot be separated by physical means*.

# Electron Configuration

Recall that an atom contains subatomic particles including:

- protons ($+$) positively charged particles
- electrons ($-$) negatively charged particles
- neutrons (0) having no charge

Free (unattached) uncharged atoms have the same number of electrons as protons, making them electrically neutral. The protons are in the nucleus and do not change or vary except in some nuclear reactions. The electrons are in discrete pathways, or shells (orbits), around the nucleus. There is a ranking or hierarchy of the shells, usually with the shells further from the nucleus having a higher energy.

In an atom, the electrons seek out the orbits that are closest to the nucleus because they are located at a lower energy level. The low-energy orbits fill first. The higher-energy levels fill with electrons only after the lower-energy levels are occupied. The lowest energy orbit is labeled as the **K shell,** which is closest to the nucleus. The outer orbits, or shells, are listed in alphabetical order: **K, L, M, N, O, P,** and **Q.**

In the atomic diagram shown in table 2.4, it can be seen that two electrons are needed to fill the K shell, eight electrons to fill the L shell, and, for light elements (atomic numbers 1–20), eight electrons will fill the M shell.

For the following discussion, refer to the Electron Configuration Chart above. *Electron configuration* is the "shape" of the electrons around the atom, that is, which energy level (shell) and what kind of orbital it is in. The shells were historically named for the chemists who found and calculated the existence of the first (inner) shells. Their names began with "K" for the first shell, then "L," then "M," so subsequent energy levels were continued up the alphabet. The numbers 1 through 7 have since been substituted for the letters.

The electron configuration is written out with the first (large) number as the shell number. The letter is the orbital type (either *s, p, d,* or *f*). The smaller superscript number is the number of electrons in that orbital.

To use this scheme, you first must know the orbitals. An *s* orbital only has two electrons. A *p* orbital has six electrons. A *d* orbital has ten electrons. An *f* orbital has fourteen electrons. We can tell what type of orbital it is by the number on the chart. The only exception to that is that "8" on the chart is "2" plus "6," that is, an *s* and a *p* orbital. The chart reads left-to-right and then down to the next line, just as English writing. Any element with over twenty electrons in the electrically neutral unattached atom will have all the electrons in the first row on the chart.

The totals on the right indicate using whole rows. If an element has an atomic number over thirty-eight, take all the first two rows and whatever more from the third row. For example, Iodine is number fifty-three on the periodic table (discussed later). For its electron configuration we would use all the electrons in the first two rows and fifteen more electrons: 1*s*2 2*s*2 2*p*6 3*p*6 4*s*2 3*d*10 4*p*6 5*s*2 from the first two rows, and 4*d*10 5*p*5 from the third row. We can add up the totals for each shell at the bottom. Full shells would give us the totals on the bottom.

**Table 2.4.    Electron Configuration Chart**

| K | L | M | N | O | P | Q | |
|---|---|---|---|---|---|---|---|
| 1 | 2 | 3 | 4 | 5 | 6 | 7 | |
| s | sp | spd | spdf | spdf | spd | sp | |
| 2 | 8 | 8 | 2 | | | | 20 |
| | | 10 | 6 | 2 | | | 38 |
| | | | 10 | 6 | 2 | | 56 |
| | | | 14 | 10 | 6 | 2 | 88 |
| | | | | 14 | 10 | 6 | |
| — | — | — | — | — | — | — | |
| 2 | 8 | 18 | 32 | 32 | 18 | 8 | TOTALS |

## Key Point!

Electron configuration is the "shape" of the electrons around an atom, that is, which energy level (shell) and what kind of orbital it is in.

# The Periodic Table of Elements

**The periodic table of elements** *is an arrangement of the elements into rows and columns in which those elements with similar properties occur in the same column.* Simply, the periodic table of elements is a way to arrange the elements, based on electronic distribution, to show a large amount of information and organization. Chemists use this table to correlate the chemical properties of known elements and predict the properties of new ones.

The periodic table of elements provides information and organization on the following:

- atomic number (the number of electrons revolving about the nucleus of the atom)
- isotopes (atoms of an element with different atomic weights)
- atomic weight and molecular weight
- groups (vertical columns) and periods (horizontal columns)
- locating important elements

The periodic table of elements provides information about the elements as shown in figure 2.4. Remember that each **element** is represented by a chemical symbol:

| | |
|---|---|
| Fe (Iron) | P (Phosphorus) |
| Al (Aluminum) | Ag (Silver) |
| Ca (Calcium) | Cl (Chlorine) |
| C (Carbon) | H (Hydrogen) |

The **atomic number** is the number of protons in the nucleus of an atom (also the number of electrons surrounding the nucleus of an atom). Remember that *protons* are positively charged subatomic particles found in the nucleus of the atom.

## Key Point!

The number of protons in the nucleus determines the atomic number. Change the number of protons and we change the element, the atomic number, and the atomic mass.

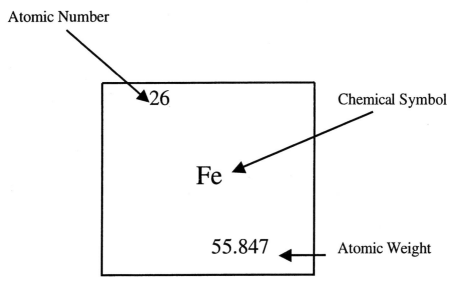

**Figure 2.4.    Element of iron.**

As mentioned, the atomic number can also indicate the *number of electrons* if the *charge* of the atom is known.

| Element | Atomic No. | Charge | No. of Electrons |
| --- | --- | --- | --- |
| O | 8 | 0 | 8 |
| $O^{-2}$ | 8 | $-2$ | 10 |
|  |  |  |  |
| Na | 11 | 0 | 11 |
| $Na^+$ | 11 | $+1$ | 10 |

The **atomic number** (again, the number of protons) is a *unique identification number* for each element. It *cannot change* without changing the identity of the element.

$$C = 6 \qquad O = 8 \qquad Cl = 17$$

The **number of neutrons** can change without changing the identity of the element or its chemical and physical properties.

*Atoms with the same number of protons but different numbers of neutrons are called* **isotopes.**

| Isotope | #P | #N |
| --- | --- | --- |
| $^1H$ (hydrogen) | 1 | 0 |
| $^2H$ (deuterium) | 1 | 1 |
| $^3H$ (tritium) | 1 | 2 |

If the number of neutrons change, the **atomic weight** of the atom changes.

| Isotope | #P | #N | Weight |
|---|---|---|---|
| $^1$H (hydrogen) | 1 | 0 | 1 au |
| $^2$H (deuterium) | 1 | 1 | 2 au |
| $^3$H (tritium) | 1 | 2 | 3 au |

**Atomic weight** *is the relative weight of an average atom of an element, based on $C^{12}$ being exactly twelve atomic units.* Chemists generally round atomic weights to the nearest whole number.

$$H = 1 \qquad O = 16 \qquad Fe = 56$$

Chemists add up atomic weights to determine the **molecular weight** of a compound. For example, the *molecular weight* of water ($H_2O$) equals the weight of 2 hydrogen atoms and 1 oxygen atom.

$$\frac{1}{2(1.008)} + \frac{16}{2(15.999)} \cong 18$$

*Molecular weights* are an important indication of the general size (and therefore complexity) of a compound. In general (but not always), the higher the molecular weight, the greater the likelihood that the compound may persist in the environment.

**Periods**, the **rows** (horizontal lines) of the periodic table, are organized by increasing atomic number (see figure 2.5). **Groups** are the **columns** of the periodic table, which contain elements with similar chemical properties. Within each **Period** (row), the *size of the nuclei increase* going from left to right because the atomic number increases (see figure 2.6).

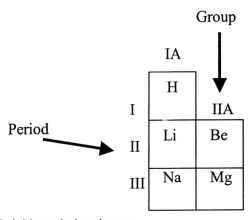

**Figure 2.5.  Periodic table period and group.**

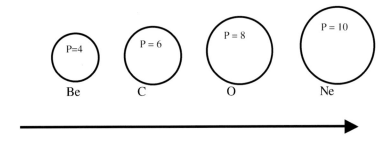

Increasing Nuclear Radius

**Figure 2.6. Increasing atomic number.**

Within each **Group** (column), the *size of the atoms increase* going from top to bottom because the number of electron shells increases. In addition, within each Group (column) elements have **similar chemical reactivity** because they have a *similar number of outer shell electrons.*

The **Group numbers** (i.e., Roman numerals) above each column indicate the number of outer-shell electrons in each Group:

Group V = 5 outer-shell electrons
Group II = 2 outer-shell electrons
Group VII = 7 outer-shell electrons

Only outer-shell electrons are involved in chemical change. The nucleus and inner shell electrons are not altered in any way during ordinary chemical reactions.

# Atomic Bonding

Now that we understand what atoms are and how they look, we can look at chemical bonds—when atoms become attached to one another. A brief review of key points sets the stage for the material that follows.

- **Matter** is anything that takes up space and has mass.
- An **element** is a substance that cannot be broken down by ordinary means—the material making up matter.
- Small units of matter are called **atoms. Protons** (+), **neutrons (0)**, and **electrons** (−) are the subunits of atoms.
- **Atomic number** is the total number of protons in an atom.
- **Atomic mass** is the total number of neutrons and protons in an atom.
- **Isotopes** are different atomic forms caused by varying the number of electrons.
- **Energy levels:** All electrons have the same mass and charge. They differ in the amounts of potential energy they possess. Electrons closer to the nucleus contain less potential energy.

- **The periodic table** is a chart made up of eighteen columns and seven rows. The elements on the table are organized by increasing atomic number. The periodicity of the elements defines unique properties such that elements of a given column in the table have similar properties.

As mentioned earlier, atoms of elements can combine to form molecules of compounds. Experimentation has shown that only certain combinations of atoms will bond together. It has also been found through experiments that the number of atoms of each element in a molecule is very definite. The best example of this is a water molecule, which is formed of two hydrogen atoms and one oxygen atom—no other combination of hydrogen and oxygen makes water. In this chapter, we provide a brief discussion of why only certain molecules occur, the different types of bonds, and chemical formulas.

## HOW ATOMS ARE LINKED TOGETHER

The tendency of elements to link together to form compounds through a shift of electronic structure is known as **valence**. This linking process is accomplished through **valence electrons**. These electrons occupy the last energy level of an atom. It is here that atoms come in contact with each other. It stands to reason that chemical bonds will occur here in any chemical reaction. The maximum number of valence electrons any atom can contain is eight. Any number less than eight will allow that atom to act as a donor or recipient of electrons to become stable. Atoms that give electrons will become positive ions and have a positive ($+$) charge, while atoms that receive electrons will become a negative ion with a negative ($-$) charge.

## ABOUT CHEMICAL BONDS

Atoms are linked or joined by **chemical bonds**. Only electrons are involved in the formation of chemical bonds between atoms, and typically only the *outermost electrons* (i.e., the *outer shell electrons*) are involved in bonding. Each bond consists of two electrons, one from each atom in the bond.

There are two different "types" of chemical bonds—depending on the type of atoms that are bonded together—**covalent** and **ionic** bonds. A **covalent bond** results from the *sharing* of a pair of electrons between two atoms. Covalent bonds are further categorized as two types, depending on how the electrons are shared:

- **Nonpolar:** Here the valence electrons are shared *equally*, thus eliminating a positive and negative end on the molecule. An example of a nonpolar bond is $CHCl_4$. It is nonpolar because although the four C-Cl bonds are all polar, the symmetry of their arrangement around the central C atom makes the overall molecule nonpolar.
- **Polar covalent bond:** Here the valence electrons are shared *unequally*, causing the molecule to develop a positive end (where the electrons spend less time) and a negative

end (where the electrons spend more time). This has to do with the electronegativity of the atom. The more electronegative the atom, the more it will hold on to the electrons. Oxygen is very electronegative. Hydrogen is not. If the difference in electronegativity between two atoms is sufficiently large, the shared electron pair will spend all of its time on the more electronegative atom, resulting in ionic bonding rather than covalent bonding. An example of a polar covalent bond is $CHCL_3$, because of the polar C-Cl bond (Cl is an electronegative atom).

## Key Point!

A polar covalent bond has a positively charged end and a negatively charged end.

In covalent bonding, the sharing of one pair of electrons is called a **single bond**; the sharing of two pairs of electrons is called a **double bond** (e.g., carbon dioxide); and the sharing of three pairs of electrons is called a **triple bond** (e.g., acetylene). Double bonds are more reactive than single bonds, and compounds containing double bonds are somewhat more volatile than corresponding single-bonded molecules. Triple bonds are even more reactive than double bonds, and volatility in triple-bonded compounds is still greater.

The second type of chemical bond, an **ionic bond**, results from the *transfer* of electrons from one atom to another, rather than two atoms sharing the electrons. In an ionic bond, one atom *donates* one or more of its outermost electrons to another atom or atoms (see figure 2.7).

- The atom that *gains* the electrons becomes a negative ion, or **anion** (Remember the mnemonic "*a n*egative *ion*").
- The atom that *loses* the electron becomes a positive ion, or **cation.**

### ELECTRONEGATIVITY AND POLARITY

**Electronegativity** is a measure of the ability of an atom to attract its outermost electrons. The higher the electronegativity, the greater the atom's ability to attract electrons. Common electronegative atoms are shown in figure 2.8.

Highly electronegative atoms tend to form *ionic* or *polar covalent* bonds. As we

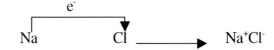

**Figure 2.7.  Ionic bond.**

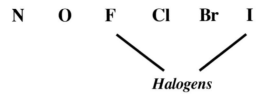

**Figure 2.8.**  Common electronegative atoms.

have seen above, in polar covalent bonds, the more electronegative atom of the two "keeps" the electron pair more (i.e., takes a larger share of the electron density). In ionic bonds, the electronegativity of one atom is so strong, that it keeps the electrons entirely to itself.

Think of chemical bonding as a continuum from nonpolar covalent bonds to ionic bonds (see figure 2.9). **Bond polarity** depends on the *electronegativity* of the atoms involved, but **overall molecular polarity** depends on symmetry (see figure 2.10).

## BOND STRENGTHS AND BOND ANGLES

Chemists measure the *strength* of a bond by determining how much energy is needed to break the bond (see figure 2.11).

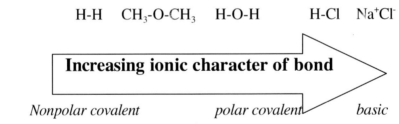

**Figure 2.9.**  Electronegativity and polarity.

**Figure 2.10.**  Bond polarity.

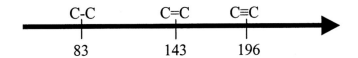

Increasing Bond Strength

(kcal/mole)

**Figure 2.11.  Bond strength.**

Other examples of Carbon (C) bond strengths are shown below.

| Type of Bond | Strength (kcal/mol) |
| --- | --- |
| C-N | 70 |
| C-Cl | 79 |
| C-O | 84 |
| C-H | 99 |

In addition to measuring bond strengths, chemists can also measure **bond angles** between atoms (see figure 2.12).

**Bond angles** depend on the type of atoms bonded together. Chemists predict bond angles and other 3D structures using computer programs (see figure 2.13).

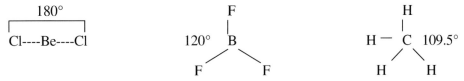

**Figure 2.12.  Bond angles.**

**Bicyclo[2.2.2.]octane**

**Cyclohexane**

**Figure 2.13.  Structure and bonding.**

# Hydrogen Bonding

**Hydrogen bonds** are bonds formed when hydrogen that is covalently bonded to an electronegative atom is attracted to another electronegative atom on another molecule. Hydrogen bonds are *weak attractions between molecules*, whereas ionic and covalent bonds are much stronger attractions between atoms. An everyday example of a hydrogen bond is in water (see figure 2.14).

When H is bonded to O or N, its lone electron is pulled away from its single proton nucleus. A broken line is used to represent hydrogen bonds because they are much weaker than covalent bonds (see figure 2.15). Hydrogen bonds are like "glue" between molecules.

Although a single **hydrogen bond** is very weak, the cumulative effect of an enormous number of **hydrogen bonds** can significantly affect the properties of the compound.

*Short Review*

When an atom combines chemically with another atom, it will either:

1. Gain electrons (become a negatively charged ion, or anion)
3. Lose electrons (become a positively charged ion, or cation)
4. Share electrons

# Crystal Structure

Smooth, angular shapes known as crystals will normally form when crystalline minerals solidify and grow without interference. The packing of atoms in a crystal structure

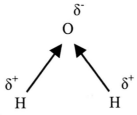

**Figure 2.14.  Hydrogen bonding.**

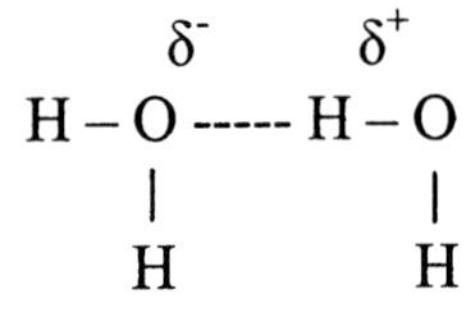

$$\delta^- \quad \delta^+$$
$$H - O ----- H - O$$
$$| \qquad\qquad |$$
$$H \qquad\qquad H$$

**Figure 2.15.  Hydrogen bond.**

requires an orderly and repeated atomic arrangement. The planes that form the outside of the crystals are known as faces. The shape of crystals and the angles between related sets of crystal faces are important in mineral identification.

Each mineral has been assigned to one of six crystal systems. The classes have been established on the basis of the number, position, and relative length of the crystal axes—imaginary lines extending through the center of the crystal.

Geologists recognize the following crystal systems:

**Isometric**—crystals belonging to this system have three axes of equal length and at right angles to one another. Examples of minerals that crystallize in the isometric system are halite, magnetite, and garnet.

**Tetragonal**—these crystals are characterized by having all three axes at right angles. The horizontal axes are of equal length but are longer or shorter than the vertical axis. Mineral species that crystallize in the tetragonal crystal system are zircon and cassiterite.

**Hexagonal**—this system has crystals marked by three horizontal axes of equal length that intersect at angles of 120 degrees, and a vertical axis at right angles to these. Example minerals that crystallize in the hexagonal division are calcite, dolomite, low quartz, and tourmaline.

**Orthorhombic**—crystals assigned to this system have all three axes all at right angles and each of different length. Minerals that belong to this system are olivine and barite.

**Monoclinic**—crystals from this system have three unequal axes, two of which intersect at right angles. Mineral species that adhere to the monoclinic crystal system include pyroxene, amphibole, orthoclase, azurite, and malachite, among others.

**Triclinic**—crystals of this system are characterized by three axes of unequal length and all oblique to one another. Mineral species of this system include plagioclase and axinite.

# Rock-Forming Minerals

Rocks consist of aggregates of minerals. Minerals are made up of one or a number of chemical elements with definite chemical compositions. Of the 3,000 or so minerals that are known to be present in the earth's crust, only about 20 are very common and relatively few (only 9) are major constituents of the more common rocks. These 9 minerals are all silicates. Those minerals that do make up a large part of the more common types of rocks are called the rock-forming minerals. Some of the more important rock-forming minerals are briefly discussed below.

**Quartz.** Quartz is the most common mineral on the face of the earth and one of the most widely distributed. It is a glassy-looking, transparent or translucent mineral that forms an important part of many igneous rocks and is common in many sedimentary and metamorphic rocks. Found in nearly every geological environment, it is at least a component of almost every rock type.

**Feldspars**. Minerals belonging to the feldspar group constitute the most important group of rock-forming minerals. Feldspar, like quartz, is a very common, light-colored rock-forming mineral. Unlike quartz, feldspar is not glassy but instead is generally dull to opaque with a porcelain-like appearance. Feldspars are found in almost all igneous rocks as well as in many sedimentary and metamorphic rocks. Feldspar is hard but can be scratched by quartz. Feldspars are split into two principal groups: orthoclase (potash feldspar) and plagioclase (soda-lime feldspar).

**Mica**. Minerals of the mica group are easily distinguished by characteristic peeling into many flat, paper-thin, smooth sheets or flakes. These sheets are chemically inert, dielectric, elastic, flexible, hydrophilic, insulating, lightweight, platy, reflective, refractive, resilient, and transparent to opaque (USGS, 2008). Mica may be white and pearly (muscovite) or dark and shiny (biotite).

**Pyroxenes**. This group is composed of complex silicates and is among the most common of all rock-forming minerals. The most common pyroxene mineral is augite, which is generally dark green to black in color. Augite is a common constituent of many of the dark-colored igneous rocks. Augite forms short, stubby crystals that have square or rectangular cross-sections. Pyroxenes are also commonly found in certain metamorphic rocks.

**Amphiboles**. The amphiboles are closely related to the pyroxenes. The most common amphibole is hornblende, a common constituent of igneous and metamorphic rocks. Hornblende is not a mineral and the name is used as a general field term. Hornblende is quite similar to augite in that both are dark; however, hornblende crystals are generally longer, thinner, and shinier than augite, and the cross-sections are diamond-shaped.

**Calcite**. Calcite (calcium carbonate) is a very common mineral in sedimentary rocks. It is commonly white to gray in color. Individual crystals are generally clear and transparent. It occurs in many sedimentary and metamorphic rocks and is the primary constituent of most limestone.

**Dolomite**. Dolomite is a common sedimentary rock-forming mineral that can be found in massive beds several hundred feet thick. It may occur in association with many ore deposits and in cavities in some igneous rocks, and it is also found in metamorphic marbles, hydrothermal veins, and replacement deposits.

**Aragonite**. Like calcite, aragonite is composed of calcium carbonate, but it differs from calcite in that it is a polymorph of calcite, which means that it has the same chemistry of calcite but a different structure. Aragonite is also less stable than calcite and crystallizes in the orthorhombic system.

**Gypsum**. Gypsum, calcium sulfate, is one of the more common minerals in sedimentary environments. It is a major rock-forming mineral that produces massive beds, usually from precipitation out of highly saline waters. Gypsum is a mineral of great economic importance; it is commonly used in the manufacture of sheet rock and plaster of Paris.

**Anhydrite**. Although chemically similar to gypsum, anhydrite does not contain water and is harder and heavier than gypsum. Anhydrite is a relatively common sedimentary mineral that forms massive rock layers.

**Halite**. Composed of sodium chloride, halite is commonly called rock salt. As a result of the evaporation of prehistoric seas, in some areas the salt occurs in thick beds (e.g., in Michigan and New York).

**Kaolinite.** Kaolinite, occurring in soft, compact, earthy masses with its dull luster and greasy feel, is typical of the other minerals that are commonly found in clay. Kaolinite is important to the production of ceramics and porcelain. The greatest demand for Kaolinite is in the paper industry to produce a glossy paper such as is used in most magazines.

**Serpentine.** Serpentine, a complex group of hydrous magnesium silicates, is a major rock-forming mineral and is found as a constituent in many metamorphic and weathered igneous rocks. It often colors many of these rocks to a green color and most rocks that have a green color probably have serpentine in some amount.

**Chlorite.** Chlorite, a complex silicate of aluminum, magnesium, and iron in combination with water, is the general name for several minerals that are difficult to distinguish by ordinary methods.

# Metallic Minerals

The metallic minerals have always been of interest to geologists because of their intrinsic value. These minerals are found in ore deposits—rock masses from which metals may be obtained commercially. Because of their importance, and using USGS (2008) descriptions, some of the more important metals and their ores are briefly discussed below.

**Aluminum.** Aluminum, the second most abundant metallic element in the earth's crust after silicon, is one of the most important metals of industry and is primarily derived from bauxite. It weighs about one-third as much as steel or copper; is malleable, ductile, and easily machined and cast; and has excellent corrosion resistance and durability.

**Copper.** Copper, one of the most useful of all metals, is usually found in nature in association with sulfur. Copper is one of the oldest metals ever used and has been one of the important materials in the development of civilization. Because of its properties of high ductility, malleability, and thermal and electrical conductivity, and its resistance to corrosion, copper has become a major industrial metal, ranking third after iron and aluminum in terms of quantities consumed.

**Gold.** Gold has been prized since the dawn of history for its great beauty and the fact that it is soft enough to be easily fashioned into coins, jewelry, and other valuable objects. Gold occurs as native gold and is typically found in quartz veins and in association with the mineral pyrite.

**Lead.** Galena, the most important source of lead, occurs in a wide variety of rocks, including igneous, sedimentary, and metamorphic. Lead is a very corrosion-resistant, dense, ductile, and malleable blue-tray metal that has been used for at least 5,000 years. Lead is used in the manufacture of paints; as typesetting metal; and for pipes, solder, metal alloys, shot, and as shielding materials to protect against radioactivity and X-rays.

**Mercury.** The most abundant ore of mercury is cinnabar (aka quicksilver), or mercuric sulfide. Mercury is the only common metal that is liquid at room temperature. Mercury has uniform volumetric thermal expansion, good electrical conductivity, and easily forms amalgams with almost all common metals except iron. Mercury is mostly used for the manufacture of industrial chemicals and for electrical and electronic applications.

**Silver.** Silver, a metal used for thousands of years as ornaments and utensils, for trade, and as the basis of many monetary systems, may occur as native silver or in one of several silver ore minerals. Silver also has many industrial applications such as in mirrors, electrical and electronic products, and photography, which is the largest single end use of silver.

**Tin.** Although widely distributed in small amounts, tin occurs in commercial quantities in igneous rocks, where it is commonly associated with quartz, topaz, galena, and tourmaline. Tin is one of the earliest metals know and used. Most tin is used as a protective coating or as an alloy with other metals such as lead or zinc.

**Iron.** Iron ore is a mineral substance that, when heated in the presence of a reductant (coke or charcoal), will yield metallic iron. Iron ore is the source of primary iron for the world's iron and steel industries. It is therefore essential for the production of steel, which in turn is essential to maintain a strong industrial base.

**Zinc.** Zinc, a rather common mineral, is a metal of considerable economic importance. It is found in veins in igneous, sedimentary, and metamorphic rocks, and as replacement deposits in limestone. Zinc is used in galvanizing steel, and in the manufacture of paint, cosmetics, types metal, dry cell batteries, and for a multitude of other purposes.

# Radioactive Minerals

Radioactive minerals have come to play an ever-increasing part in modern technology. Some of these minerals are widely used as sources of energy for nuclear power plants, in nuclear medicine, and in modern weapon systems. Examples of common radioactive minerals include autunite, brannerite, carnotite, monazite, thorianite, and uraninite. The vast majority of the radioactive content in minerals or ores is either uranium-238 or thorium-232.

# Nonmetallic Minerals

Nonmetallic minerals refer to a vast category that includes coal, petroleum, sulfur, fertilizer, building stones, and gemstones. Industrial minerals include, for example, flake graphite, silicon carbide, garnet, talc, marble, slate, granite, salt, sulfur, asbestos, and others.

# Chapter Review Questions

2.1    Define matter.

2.2    Mineral color, taste, melting point, boiling point, density, luster, and hardness are _____ physical properties.

2.3    _____ is the mass per unit volume of a substance.

2.4    _____ is a unitless measure of the weight of a substance compared to the weight of an equal volume of water.

2.5    The appearance of the surface of a mineral as seen in reflected light is called _____.

2.6    _____ are the basic building blocks of all matter.

2.7    _____ are the smallest units of a compound that still retain the properties of that compound.

2.8    _____ is the total number of neutrons and protons in an atom.

2.9    Atoms are linked by _____.

2.10  A positive ion is a _____.

# References and Recommended Reading

Asimov, I., and Back, D. F. (Illustrator). 1992. *Atom: Journey across the Subatomic Cosmos*. New York: Dutton/Plume.

Levi, Rimor, Rosenthal R. (Translator). 1986. *The Periodic Table (American)*. New York: Knopf Publishing Group.

Mebane, R. C., and Rybolt, T. R. 1998. *Adventures with Atoms and Molecules: Chemistry Experiments for People*, Vol. 1. Hillside, NJ: Enslow Publishers.

Nardo, D. 2001. *Atoms*. New York: Gale Group.

Senese, F. 2004. *What Is Chemistry?* Accessed 05/25/08 at http://antoine.frostburg.edu/chem/senese/101/intro/faq/what-is-ch emistry.shtml.

Spellman, F. R. 2006. *Chemistry for Non-Chemists*. Lanham, MD: Government Institutes Press.

Stwertka, A. 2002. *Guide to the Elements*. New York: Oxford University Press.

USGS. 2008. *Minerals Information*. Accessed 05/26/08 at http://minerals.usgs.gov/mineral/index.html.

# Note

1. For an easy-to-understand introduction to general chemistry, read F. R. Spellman, *Chemistry for Non-Chemists* (Lanham, MD: Government Institutes Press, 2006).

## CHAPTER 3

# Igneous Rocks and Magma Eruption

Igneous (Latin: *ignis*, "fire") rocks are those rocks that have solidified from an original molten silicate state. The occurrence and distribution of igneous rocks and igneous rock types can be related to the operation of plate tectonics. The molten rock material from which igneous rocks form is called magma. Magma, characterized by a wide range of chemical compositions and with high temperature, is a mixture of liquid rock, crystals, and gas. Magmas are large bodies of molten rock deeply buried within the earth. These magmas are less dense than surrounding rocks, and will therefore move upward. In the upward movement, sometimes magmatic materials are poured out upon the surface of the earth as, for example, when lava flows from a volcano. These igneous rocks are volcanic or **extrusive rocks**; they form when the magma cools and crystallizes on the surface of the earth. Under certain other conditions, magma does not make it to the earth's surface and cools and crystallizes within the earth's crust. These intruding rock materials harden and form **intrusive** or **plutonic rocks**.

## Magma

Magma is molten silicate material and may include already formed crystals and dissolved gases. The term *magma* applies to silicate melts within the earth's crust. When magma reaches the surface it is referred to as lava. The chemical composition of magma is controlled by the abundance of elements in the earth. These include oxygen, silicon, aluminum, hydrogen, sodium, calcium, iron, potassium, and manganese, which make up 99 percent. Because oxygen is so abundant, chemical analyses are usually given in terms of oxides. Silicon dioxide ($SiO_2$, also known as silica) is the most abundant oxide. Because magma gas expands as pressure is reduced, magmas have an explosive character. The flow (or viscosity) of magma depends on temperature, composition, and gas content. Higher silicon dioxide content and lower temperature magmas have higher viscosity.

Magma consists of three types: basaltic, andesitic, and rhyolitic. Table 3.1 summarizes the characteristics of each type.

## Intrusive Rocks

Intrusive (or plutonic rocks) are rocks that have solidified from molten mineral mixtures beneath the surface of the earth. Intrusive rocks that are deeply buried tend to

**Table 3.1   Characteristics of Magma Types**

| Magma Type | Solidified Volcanic | Solidified Plutonic | Chemical Composition | Temperature |
|---|---|---|---|---|
| Basaltic | Basalt | Gabbro | 45–55% silicon dioxide | 1,000–1,200°C |
| Andesitic | Andesite | Diorite | 55–65% silicon dioxide | 800–1,000°C |
| Rhyolitic | Rhyolite | Granite | 65–75% silicon dioxide | 650–800°C |

cool slowly and develop a coarse texture. On the other hand, those intrusive rocks near the surface that cool more quickly are finer textured. The shape, size, and arrangement of the grains comprising it determine the texture of igneous rocks. Because of crowded conditions under which mineral particles are formed, they are usually angular and irregular in outline. Typical intrusive rocks include

- **gabbro**—a heavy, dark-colored igneous rock consisting of coarse grains of feldspar and augite
- **peridotite**—a rock in which the dark minerals are predominant
- **granite**—the most common and best known of the coarse-textured intrusive rocks
- **syenite**—resembles granite, but is less common in its occurrence and contains little or no quartz

## Extrusive Rocks

Extrusive (or volcanic) rocks are those that pour out of craters of volcanoes or from great fissures or cracks in the earth's crust and make it to the surface of the earth in a molten state (liquid lava). Extrusive rocks tend to cool quickly and typically have small crystals (because fast cooling does not allow large crystals to grow). Some cool so rapidly that no crystallization occurs, and this produces volcanic glass.

Some of the more common extrusive rocks are felsite, pumice, basalt, and obsidian.

- **felsite**—very fine textured igneous rocks
- **pumice**—frothy lava that solidifies while steam and other gases bubble out of it
- **basalt**—world's most abundant fine-grained extrusive rock
- **obsidian**—volcanic glass; cools so fast that there is no formation of separate mineral crystals

## Bowen's Reaction Series

The geologist Norman L. Bowen, back in the early 1900s, was able to explain why certain types of minerals tend to be found together while others are almost never associated with one another. Bowen found that minerals tend to form in specific sequences in igneous rocks, and these sequences could be assembled into a composite sequence. The idealized progression he determined is still accepted as the general model (see figure 3.1) for the evolution of magmas during the cooling process.

In order to better understand Bowen's Reaction Series, it is important to define key terms:

**magma**—molten igneous rock
**felsic**—white pumice
**pumice**—textured form of volcanic rock; a solidified frothy lava
**extrusion**—magma intruded or emplaced beneath the surface of the earth
**feldspar**—the family of minerals including microcline, orthoclase, and plagioclase
**mafic**—a mineral containing iron and magnesium
**aphanitic**—mineral grains too small to be seen without a magnifying glass
**phaneritic**—mineral grains large enough to be seen without a magnifying glass
**reaction series**—a series of mineral in which a mineral reacts to change to another
**rock-forming mineral**—the minerals commonly found in rocks. Bowen's Reaction
    Series lists all of the common ones in igneous rocks
**specific gravity**—the relative mass or weight of a material compared to the mass or
    weight of an equal volume of water

Some igneous rocks are named according to textural criteria:

• scoria—porous
• pumice—foam

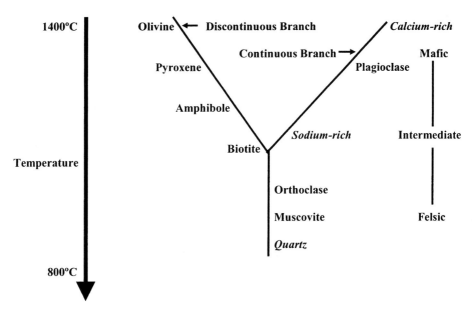

**Figure 3.1.   Bowen's Reaction Series.**

*Source*: Adapted from GeoMan (2008), http://jersey.uorgeon.edu/~mstrick/AskGeoMan/geoQuerry 32.html.

- obsidian—glass
- tuff—cemented ash
- breccia—cemented fragments
- permatite—extremely large crystals
- aplite—sugary texture, quartz and feldspar
- porphyry—fine matrix, large crystals

## THE DISCONTINUOUS REACTION SERIES

The left side of figure 3.1 shows a group of mafic or iron-magnesium bearing minerals—olivine, pyroxene, amphibole, and biotite. If the chemistry of the melt is just right, these minerals react discontinuously to form the next mineral in the series. If there is enough silica in the igneous magma melt, each mineral will change to the next mineral lower in the series as the temperature drops. Descending down Bowen's Reaction Series, the minerals increase in the proportions of silica in their composition. In basaltic melt, as shown in figure 3.1, olivine will be the first mafic mineral (silicate mineral rich in magnesium and iron) to form. When the temperature is low enough to form pyroxene, all of the olivine will react with the melt to form pyroxene, and pyroxene will crystallize out of the melt. At the crystallization temperature of amphibole, all the pyroxene will react with the melt to form amphibole and amphibole will crystallize. At the crystallization temperature of biotite, all of the amphibole will react to form biotite and biotite will crystallize. Thus all igneous rocks should only have biotite; however, this is not the case. In crystallizing olivine, if there is not enough silica to form pyroxene, then the reaction will not occur and olivine will remain. Additionally, in crystallizing olivine if the temperature drops too fast for the reaction to take place (volcanic magma eruption), then the reaction will not have time to occur, the rock will solidify quickly, and the mineral will remain olivine.

## THE CONTINUOUS REACTION SERIES

The right side of figure 3.1 shows the plagioclases. Plagioclase minerals have the formula $(Ca, Na)(Al, Si)_3O_8$. The highest temperature plagioclase has only calcium (Ca). The lowest temperature plagioclase has only sodium (Na). In between, these ions mix in a continuous series from 100 percent Ca and 0percent Na at the highest temperature to 50 percent Ca and 50 percent Na at the middle temperature to 0 percent Ca and 100 percent Na at the lowest temperature. In a basaltic melt, for example, the first plagioclase to form could be 100 percent Ca and 0 percent Na plagioclase. As the temperature drops, the crystal reacts with the melt to form 99 percent Ca and 1 percent Na plagioclase, and 99 percent Ca and 1 percent Na plagioclase crystallizes. Then those react to form 98 percent Ca and 2 percent Na, and this composition would crystallize, and so forth. All of this happens continuously provided there is enough time for the reaction s to take place and enough sodium, aluminum, and silica in the melt to form each new mineral. The end result will be a rock with plagioclases with the same ratio of Ca to Na as the starting magma.

## Key Point!

In regards to the Bowen Reaction Series, on both sides of the reaction series shown in figure 3.1, the silica content of the minerals increases as the crystallization trend heads downward. Biotite has more silica than olivine. Sodium plagioclase has more silica than calcium plagioclase.

## Important Point!

The magma temperature and the chemical composition of the magma determine what minerals crystallize and thus what kind of igneous rock we get.

# Eruption of Magma

The volcanic processes that lead to the deposition of extrusive igneous rocks can be studied in action today, and help us to explain the textures of ancient rocks with respect to depositional processes. Some of the major features of volcanic processes and landforms are discussed in this section. The information below is from USGS (2008) *Principal Types of Volcanoes*.

## TYPES OF VOLCANOES

Geologists generally group volcanoes into four main kinds—cinder cones, composite volcanoes, shield volcanoes, and lava domes.

- **Cinder cones**—are the simplest type of volcano. They are built from particles and blobs of congealed lava ejected from a single vent. As the gas-charged lava is blown violently into the air, it breaks into small fragments that solidify and fall as cinders around the vent to form a circular or oval cone. Most cinder cones have a bowl-shaped crater at the summit and rarely rise more than a thousand feet or so above their surroundings. Cinder cones are numerous in western North America as well as throughout other volcanic terrains of the world.
- **Composite volcanoes**—are some of earth's grandest mountains, sometimes called stratovolcanoes. They are typically steep-sided, symmetrical cones of large dimension

built of alternating layers of lava flows, volcanic ash, cinders, blocks, and bombs and may rise as much as 8,000 feet above their bases. Most composite volcanoes have a crater at the summit that contains a central vent or a clustered group of vents. Lavas either flow through breaks in the crater wall or issue from fissures on the flanks of the cone. Lava, solidified within the fissures, forms dikes that act as ribs and greatly strengthen the cone. The essential feature of a composite volcano is a conduit system through which magma from a reservoir deep in the earth's crust rises to the surface. The volcano is built up by the accumulation of material erupted through the conduit and increases in size as lava, cinders, ash, etc., are added to its slopes.

- **Shield volcanoes**—are built almost entirely of fluid lava flows. Flow after flow pours out in all directions from a central summit vent, or group of vents, building a broad, gently sloping cone of flat, domical shape, with a profile much like that of a warrior's shield. They are built up slowly by the accretion of thousands of highly fluid lava flows called basalt lava that spread widely over great distances and then cool as thin, gently dipping sheets. Lava also commonly erupts from vents along fractures (rift zones) that develop on the flanks of the cone. Some of the largest volcanoes in the world are shield volcanoes (see figure 3.2).

- **Lava domes**—are formed by relatively small bulbous masses of lava too viscous to flow any great distance; consequently, on extrusion, the lava piles over and around its vent. A dome grows largely by expansion from within. As it grows, its outer surface cools and hardens, then shatters, spilling loose fragments down its sides. Some domes form craggy knobs or spines over the volcanic vent, whereas others form short, steep-sided lava flows known as "coulees" (Fr. *couler*, "to flow"). Volcanic domes commonly occur within the craters or on the flanks of large composite volcanoes.

## TYPES OF VOLCANIC ERUPTIONS

The type of volcanic eruption is often labeled with the name of a well-known volcano where characteristic behavior is similar—hence the use of such terms as "Strombolian," "Vulcanian," "Vesuvian," "Pelean," "Hawaiian," and others.

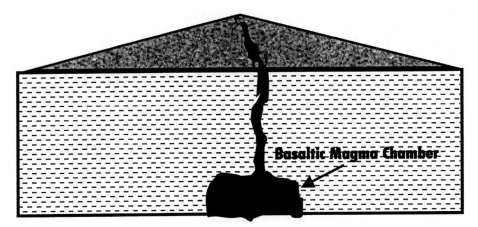

**Figure 3.2.   Cross-section of a shield volcano.**

- **Strombolian-type eruption**—this type of eruption is in constant action, with huge clots of molten lava bursting from the summit crater to form luminous arcs through the sky. Collecting on the flanks of the cone, lava clots combine to steam down the slopes in fiery rivulets.
- **Vulcanian-type eruption**—this type of eruption is characterized by very viscous lavas; a dense cloud of ash-laden gas explodes from the crater and rises high above the peak. Steaming ash forms a whitish cloud near the upper level of the cone.
- **Pelean-type eruption (or "nuee ardente"—glowing cloud)**—this type of eruption is characterized by its explosiveness. It erupts from a central crater with violent explosions that eject great quantities of gas, volcanic ash, dust, incandescent lava fragments, and large rock fragments.
- **Hawaiian-type eruption (or quiet)**—is characterized by less viscous lavas that permit the escape of gas with a minimum of explosive violence. In fissure-type eruptions, a fountain of fiery lava erupts to a height of several hundred or more feet. Such lava may collect in old pit craters to form lava lakes, or form cones, or feed radiating flows.
- **Vesuvian eruption**—is characterized by great quantities of ash-laden gas that are violently discharged to form cauliflower-shaped clouds high above the volcano.
- **Phreatic (or steam-blast) eruption**—is driven by explosive expanding steam resulting from cold ground or surface water coming into contact with hot rock or magma. The distinguishing feature of phreatic explosions is that they only blast out fragments of preexisting solid rock from the volcanic conduit; no new magma is erupted.
- **Plinian eruption**—is a large explosive event that forms enormous dark columns of tephra (solid material ejected) and gas high into the stratosphere. Such eruptions are named for Pliny the Younger, who carefully described the disastrous eruption of Vesuvius in AD 79. This eruption generated a huge column of tephra into the fall. Many thousands of people evacuated areas around the volcano, but about 2,000 were killed, including Pliny the Elder.

## Types of Lava

The two Hawaiian words *pahoehoe* and *aa*, are used to describe how lava flows. Pahoehoe (Pa-hoy-hoy) is the name for smooth or ropy lava. Cooler lava hardens on the surface; hotter, more fluid lava flows under it, often leaving caves or tubes behind.

Aa (ah-ah) is the name of rough, jagged, lava. The molten lava is much less fluid and usually moves slower. A crust never hardens on the surface, but chunks of cooler rock tumble along the top and sides instead. Aa can be impassable.

# Lava Flow Terminology

This information comes from *Photo Glossary of Volcano Terms* (USGS, 2000).

**Lava flow**—a mass of molten rock that pours onto the earth's surface during an effusive eruption. Both moving lava and the resulting solidified deposit are referred to as lava flows. Because of the wide range in (1) viscosity of the different lava types (basalt, andesite, dacite, and rhyolite); (2) lava discharge during eruptions; and (3) characteristics of the erupting vent and topography over which lava travels, lava flows come in a great variety of shapes and sizes.

**Lava cascade**—a cascade of water is a small waterfall formed as water descends over rocks. In similar fashion, a lava cascade refers to the rush of lava descending over a cliff. In Hawaii, lava cascades typically occur when lava spills over the edge of a crater, a fault scarp, or a sea cliff into the ocean.

**Lava drapery**—is the cooled, congealed rock on the face of a cliff, crater, or fissure formed by lava pouring or cascading over their edges.

**Lava channels**—are narrow, curved, or straight open pathways through which lava moves on the surface of a volcano. The volume of lava moving down a channel fluctuates, so the channel may be full or overflowing at times and nearly empty at other times. During overflow, some of the lava congeals and cools along the banks to form natural levees that may eventually enable the lava channel to build a few meters above the surrounding ground.

**Standing waves**—in a fast-moving lava flow appear to be stationary relative to the lava that moves over the land through them, similar to the standing waves in a water stream. In Hawaii, standing waves as high as three meters have been observed.

**Lava spillways**—are confined lava channels on the sides of a volcanic cone or shield that form when lava overflows the rim of the vent.

**Lava surge**—intermittent surges or accelerations in the forward advance of lava can occur when the supply of lava to a flow suddenly increases or a flow front gives way. The supply of lava may increase as consequence of a higher discharge of lava from the vent, a sudden change in the vent geometry so that a great volume of lava escapes (e.g., the collapse of a vent wall), or by the escape of ponded lava from along a channel. Lava surges may be accompanied by thin, short-lived breakouts of fluid lava from the main channel and flow front.

**Methane explosion**—sudden explosions of methane gas occur frequently near the edges of active lava flows. Methane gas is generated when vegetation is covered and heated by molten lava. The explosive gas travels beneath the ground through cracks and fills abandoned lava tubes for long distances around the margins of the flow. Methane gas explosions have occurred at least 100 meters from the leading edge of a flow, blasting rocks and debris in all directs.

**Volcanic domes**—are rounded, steep-sided mounds built by very viscous magma, usually either dacite or rhyolite. Such magmas are typically too viscous (resistant to flow) to move far from the vent before cooling and crystallizing. Domes may consist of one of more individual lava flows. Volcanic domes are also referred to as lava domes.

## Did You Know?

The longest historical dome-building eruption is still occurring at Santiaguito Dome, which is erupting on the southeast flank of Santa Maria volcano in Guatemala; the dome began erupting in 1922.

# Intrusions

Intrusive (or plutonic igneous) rocks have been intruded or injected into the surrounding rocks. Some of these intrusions are invisible because they are embedded at great depth; consequently, igneous intrusive bodies may be seen only after the underlying rocks have been removed by erosion.

Intrusions are of two types: **Concordant** intrusions, which are parallel to layers of rocks, and **discordant** intrusions, which cut across layers. Some of the more common intrusive bodies (plutons) are discussed below.

## CONCORDANT INTRUSIONS

- **Sills**—are tabular bodies of igneous rocks that spread out as essentially thin, horizontal sheets between beds or layers of rocks.
- **Laccoliths**—are lenslike, mushroom-shaped, or blisterlike intrusive bodies, usually near the surface, that have relatively flat under surfaces and arched or domed upper surfaces. They differ from sills in that they are thicker in the center and become thinner near their margins.
- **Lopoliths**—are mega-sills, usually of gabbro or diorite, that may cover hundreds of square kilometers and be kilometers thick. They often have a concave structure and are differentiated. That is, they take so long to harden that heavy minerals have a chance to sink and light minerals can rise.

## DISCORDANT INTRUSIONS

- **Dikes**—are thin, wall-like sheets of magma intruded into fractures in the crust.
- **Stocks** or **plutons**—are small irregular intrusions.
- **Batholiths**—are the largest of igneous intrusions, usually granitic, and cover hundreds or thousands or square kilometers.

## Did You Know?

Most obsidian is black, but red, green, and brown obsidian are known. Obsidian forms when magma is cooled so quickly that individual minerals cannot crystallize.

# Volcanic Landforms

Volcanic landforms (or volcanic edifices) are controlled by the geological processes that form them and act on them after they have formed. There are four principal types of volcanic landforms: plateau basalts or lava plains, volcanic mountains, craters, and calderas.

- **Plateau basalts and lava plains**. These are formed when great floods of lava are released by fissure eruptions instead of central vents and spread in sheetlike layers over the earth's surface, forming broad plateaus. Some of these plateaus are quite extensive. For example, the Columbia River Plateau of Oregon, Washington, Nevada, and Idaho is covered by 200,000 square miles of basaltic lava.
- **Volcanic mountains**. These are mountains that are composed of the volcanic products of central eruptions. They are classified as cinder cones (conical hills), composite cones (stratovolcanoes), and lava domes (shield volcanoes).
- **Volcanic craters**. These are circular, funnel-shaped depressions, usually less than one kilometer in diameter, that form as a result of explosions that emit gases and tephra
- **Calderas**. These are much larger depressions, circular to elliptical in shape, with diameters ranging from one to fifty kilometers. Calderas form as a result of collapse of a volcanic structure. The collapse results from evacuation of the underlying magma chamber.

# Thermal Areas

Thermal areas are locations where volcanic or other igneous activity takes place, as is evidenced by the presence and/or action of volcanic gases, steam, or hot water escaping from the earth.

- **Fumaroles**. These are vents where gases, either from a magma body at depth, or steam from heated groundwater, emerge at the surface of the earth.
- **Hot springs**. These hot springs or thermal springs are areas where hot water comes to the surface of the earth. Cool groundwater moves downward and is heated by a body of magma or hot rock. A hot spring results if this hot water can find its way back to the surface, usually along fault zones.
- **Geysers**. A geyser results if the hot spring has a plumbing system that allows for the

accumulation of steam form the boiling water. When the steam pressure builds so that it is higher than the pressure of the overlying water in the system, the steam will move rapidly toward the surface, causing the eruption of the overlying water. Some geysers, like Old Faithful in Yellowstone Park, erupt at regular intervals, but most geysers are quite erratic in their performance. The time between eruptions is controlled by the time it takes for the steam pressure to build in the underlying plumbing system.

# Chapter Review Questions

3.1 _____ are large bodies of molten rock deeply buried with the earth.

3.2 Magma consists of three types: _____, _____, and _____.

3.3 _____ is a heavy, dark-colored igneous rock consisting of coarse grains of feldspar and augite.

3.4 _____ is a finely textured igneous rock.

3.5 The _____ is the simplest type of volcano.

3.6 _____ are confined lava channels on the sides of a volcanic cone or shield that form when lava overflows the rim of the vent.

3.7 _____ have been intruded or injected into the surrounding rocks.

3.8 _____ are tabular bodies of igneous rocks that spread out as essentially thin, horizontal sheets between beds or layers of rocks.

3.9 _____ are small irregular intrusions.

3.10 Magma body gas vents: _____.

# References and Recommended Reading

Abbott, P. L. 1996. *Natural Disasters*. New York: Wm. C. Brown.

Anderson, J. G., Bodin, P., Brune, H. N., Prince, J., Singh, S. K., Quaas, R., and Onate, M. 1986. Strong ground motion from the Michoacan, Mexico, earthquake. *Science* 233, pp. 1043–49.

Browning, J. M. 1973. Catastrophic rock slides. Mount Huascaran, north-central, Peru, May 32, 1970. *Bulletin American Association of Petroleum Geologists* 57, pp. 1335–1341.

Coch, N. K. 1995. *Geohazards, Natural and Human*. New York: Prentice Hall.

Eagleman, J. 1983. *Severe and Unusual Weather*. New York: Van Nostrand Reinhold.

Francis, P. 1993. *Volcanoes: A Planetary Perspective*. New York: Oxford University Press.

GeoMan. 2008. *Bowen's Reaction Series*. Accessed at http://jersey.uoregon.edu/~mstrick/Ask GeoMan/geoQuerry32.html.

Keller, E. A. 1985, *Environmental Geology*, 4th ed. New York: Merrill.

Kiersh, G. A. 1964. Vaiont reservoir disaster. *Civil Engineering* 34, pp. 32–39.

Murck, B. W., Skinner, B. J., and Porter, S. C. 1997. *Dangerous Earth: An Introduction to Geologic Hazards*. New York: Wiley.

Skinner, B. J., and Porter, S. C. 1995. *The Dynamic Earth: An Introduction to Physical Geology*, 3rd ed. New York: Wiley.

Spellman, F. R., and Whiting, N. E. 2006. *Environmental Science and Technology*, 2nd ed. Rockville, MD: Government Institutes Press.

Stephens, J. C., et al. 1984, Organic soils subsidence. *Geological Society of American Reviews in Engineering Geology* 6, p. 3.

Swanson, D. A., Wright, T. H., and Helz, R. T. 1975. Linear vent systems and estimated rates of magma production and eruption of the Yakima basalt on the Columbia Plateau. *American Journal of Science* 275, pp. 877–905.

Tilling, R. I. 1984. *Eruptions of Mount St. Helens: Past, Present and Future.* Department of the Interior, U.S. Geological Survey.

USGS. 1989. Lessons learned from the Loma Prieta, California, earthquake of October 17, 1989. circular 1045.

USGS. 2000. *Photo Glossary of Volcano Terms.* Accessed 06/01/08 at http://volcanoes.usgs.gov/images/pglossary/index.php.

USGS. 2008. *Principal Types of Volcanoes.* Accessed 05/31/08 at http://pubs.usgs.gov/gip/volc/types.html.

Williams, H., and McKinney, A. R. 1979, *Volcanology.* New York: Freeman & Copper Co.

CHAPTER 4

# Sedimentary Rocks

A sedimentary rock is like a history book.

—USGS (2006)

When rocks are exposed on the earth's surface, they are especially vulnerable to the surface agents of erosion (weathering, rain, streamflow, wind, wave action, ocean circulation). When eroded, these rock fragments (called **detritus**) are commonly picked up and transported by wind, water, and ice, and when the transporting agent has dropped them (transporting energy is not strong enough to carry them), they are generally referred to as **sediments**. Sediments are composed of small fragments (gravel, sand, or silt size); new minerals (mainly clays); and dissolved portions the source rock (dissolved salts in river and ocean water). Sediments on the earth's surface may form by mere mechanical accumulation (wind or water) such as gravel and sand deposits in a river or sand dunes in a desert; by chemical precipitation, such as salt and calcite precipitation in shallow seas and lakes; and by activity of organisms, such as carbonate accumulation in coral reefs or accumulation of organic matter in swamps (coal precursor). Sediments are typically deposited in layers or beds called **strata**. When sediments become compacted and cemented together (a process known as **lithification**), they form sedimentary rocks.

This compaction, or lithification, of sedimentary materials into stratified layers is probably the most significant feature of sedimentary rocks. These stratified layers are like pages in the ultimate history book—earth's history—with each page dedicated to a specific or particular time frame (earliest to present). Sediments of any particular time period form a distinct layer that is underlain and overlain by equally distinct layers of respectively older and younger times. These layers, composed of such common rock types as sandstone, shale, and limestone, make up about 75 percent of the rocks exposed on the earth's surface. Geologists can study sedimentary rocks in the making; therefore, they probably know more details about the origin of these rocks than that of igneous and metamorphic rocks combined.

## Types of Sedimentary Rocks

Several different types of sedimentary rocks can be distinguished according to the source of rock materials that form them. The main groupings are:

**Clastic (or detrital) sedimentary rocks**, subdivided into
- conglomerates
- sandstones
- mudstones/shales

**Chemical and biochemical sedimentary rocks**, subdivided into
- limestone/dolostone
- evaporites
- carbonaceous rocks

## CLASTIC SEDIMENTARY ROCKS

Clastic sedimentary rocks are the rocks most people think of when they think of sedimentary rocks. Clastic sedimentary rocks are made up of pieces (fragmented material from other rocks—clasts) of preexisting rocks. Pieces of rock are loosened by weathering, and then transported (water, wind, gravity, glacial action) to some basin or depression where sediment is trapped. If the sediment is buried deeply, it becomes compacted and cemented, forming sedimentary rock. Depending on grain size, they are subdivided into conglomerate, sandstone, siltstone, and shale (see table 4.1).

The formation of a clastic sedimentary rock involves the following processes:

- **Transportation**—sediments move to their final destination by sliding down slopes, being picked up by the wind, or by being carried by running water in streams, rivers, or ocean currents. During transport, the sediment particles will be sorted according to size and density and will be rounded by abrasion. The distance the sediment is transported and the energy of the transporting medium all leave clues in the final sediment that tell us something about the mode of transportation.
- **Deposition**—sediment is deposited when the energy of the transporting medium becomes too low to continue the transport process. In the deposition process, when the velocity of the transporting medium becomes too low to transport sediment, the sediment will fall out and become deposited.
- **Diagenesis**—is the process of chemical and physical change that turns sediment into rock. The first step in the process is compaction, which occurs when the weight of the overlying material increases. As the grains of the material are compacted together, pore

**Table 4.1.   Classification of Clastic Sedimentary Rock**

| Name of Particle | Size Range | Loose Sediment | Consolidated Rock |
|---|---|---|---|
| Boulder | >256 mm | Gravel | Conglomerate |
| Cobble | 64–256 mm | Gravel | Conglomerate |
| Pebble | 2–64 mm | Gravel | Conglomerate |
| Sand | 1/16–2 mm | Sand | Sandstone |
| Silt | 1/256–1/16 mm | Silt | Siltstone |
| Clay & shale | <1/256 mm | Clay | Claystone, mudstone, |

*Source*: Adapted from Fichter (2000).

space is reduced and water is eliminated from the substance. The free water usually carries mineral components in solution, and these constituents may later precipitate as new minerals in the pore spaces. This causes cementation, which will then start to bind the individual particles together and can be seen in quartz, calcite, iron oxide, clay, glauconite, and feldspar. The next stage of diagenesis involves alteration. Consider limestone and plagioclase, for example, both of these primary rocks can be converted or altered, respectively, to dolomite and to albite. This alteration occurs through pressure solution, which occurs when carbonate rock begins to dissolve under pressure, either of deep burial or tectonic squeezing. Finally, limestone is precipitated and formed in the recrystallization process. In addition, an absence of oxygen during the compaction process may cause other alterations to the original sediment.

Some of the more common types of clastic sedimentary rocks are described below.

**Shale.** Shale consists of consolidated clay and other fine particles (mud) that have hardened into rock. It is the most abundant of all sedimentary rocks, comprising about 60–70 percent of the sedimentary rocks on earth. Characteristically fine grained and thinly bedded, shale is split easily along dividing (bedding) planes. Shale is classified or typed by composition. For example, shale containing large amounts of clay is called *argillaceous*. Shale containing appreciable amounts of sand is called *arenaceous* shale. Shale high in organic matter is typically black in color and known as *carbonaceous* shale. Shale that contains large amounts of lime is known as *calcareous shale*, used in the manufacture of Portland cement. Another type of shale, *oil* shale, is currently of great interest worldwide because of (at the time of this writing) the supply and demand and increasing cost of crude oil—oil shale (which contains fossilized insoluble organic materials called *kerogen*, which is converted into petroleum products) may be a short-term solution to crude oil shortage problems (see Case Study 4.1).

**Sandstone.** Sandstones, composed essentially of cemented sand, comprise about 30 percent of all sedimentary rocks. The most abundant mineral in sandstone is quartz, along with lesser amounts of calcite, gypsum, and various iron compounds. Sandstone is used as an abrasive (for sandpaper) and as a building stone.

**Conglomerate.** These are consolidated gravel deposits with variable amounts of sand and mud between the pebbles and are the least abundant sediment type. Conglomerates accumulate in stream channels, along the margins of mountain ranges, and on beaches. Conglomerates composed largely of angular pebbles are called *breccias*; those formed in glacial deposits are called *tillites*.

## Case Study 4.1—U.S. Oil Shale Deposits[1]

Oil shale is commonly defined as a fine-grained sedimentary rock containing organic matter that yields substantial amounts of oil and combustible gas upon destructive distillation. Most of the organic matter is insoluble in ordinary organic solvents; therefore, it must be decomposed by heating to release such materials. Underlying most definitions of oil shale is its potential for the economic recovery of energy, including shale oil and combustible gas, as well as a number of by-products. A deposit of soil shale having economic potential is generally one that is at or near enough to the surface to be developed by open-pit or conventional underground mining or by in situ methods.

Oil shales range widely in organic content and oil yield. Commercial grades of oil shale, as determined by their yield of shale oil, range from about 100 to 200 liters per metric ton (l/t) of rock. The U.S. Geological Survey has used a lower limit of about 40 l/t for classification of federal oil shale lands. Others have suggested a limit as low as 25 l/t.

Deposits of oil shale are in many parts of the world. These deposits, which range from Cambrian to Tertiary age, may occur as minor accumulations of little or no economic value or giant deposits that occupy thousands of square kilometers and reach thicknesses of 700 meters or more. Oil shales were deposited in a variety of depositional environments, including freshwater to highly saline lakes, epicontinental marine basins and subtidal shelves, and in limnic and coastal swamps, commonly in association with deposits of coal.

In terms of mineral and elemental content, oil shale differs from coal in several distinct ways. Oil shales typically contain much larger amounts of inert mineral matter (60–90 percent) than coals, which have been defined as containing less than 40 percent mineral matter. The organic matter of oil shale, which is the source of liquid and gaseous hydrocarbons, typically has higher hydrogen and lower oxygen content than that of lignite and bituminous coal.

In general, the precursors of the organic matter in oil shale and coal also differ. Much of the organic matter in oil shale is of algal origin, but may also include remains of vascular land plants that more commonly compose much of the organic matter in coal. The origin of some of the organic matter in oil shale is obscure because of the lack of recognizable biologic structures that would help identify the precursor organism. Such materials may be of bacterial origin or the product of bacterial degradation of algae or other organic matter.

The mineral component of some oil shales is composed of carbonates including calcite, dolomite, and siderite, with lesser amounts of aluminosilicates. For other oil shales, the reverse is true—silicates including quartz, feldspar, and clay minerals are dominant and carbonates are a minor component. Many oil shale deposits contain small, but ubiquitous, amounts of sulfides including pyrite and marcasite, indicating that the sediments probably accumulated in dysaerobic to anoxic waters that prevented the destruction of the organic matter by burrowing organisms and oxidation.

In the past, shale oil in the world market was not competitive with petroleum, natural gas, or coal; however, this may change if the price of oil continues to climb. Even before shale oil began to become somewhat competitive with current world market prices for crude oil, it was being used in several countries that possess easily exploitable deposits of oil shale but lack other fossil fuel resources. Some oil shale deposits contain minerals and metals that add by-product value such as alum, nahcolite (sodium bicarbonate), dawsonite (sodium aluminum carbonate hydroxide), sulfur, ammonium sulfate, vanadium, zinc, copper, and uranium.

The gross heating value of oil shales on a dry-weight basis ranges form about 500 to 4,000 kilocalories per kilogram (kcal/kg) of rock. The higher-grade kukersite oil shale of Estonia, which fuels several electric power plants, has a heating value of about 2,000 to 2,200 kcal/kg. By comparison, the heating value of lignitic cola ranges form 3,500 to 4,600 kcal/kg on a dry, mineral-free basis (ASTM, 1966).

Tectonic events and volcanism have altered some deposits. Structural deformation may impair the mining of an oil shale deposit, whereas igneous intrusions may have thermally degraded the organic matter. Thermal alteration of this type may be restricted to a small part of the deposit, or it may be widespread, making most of the deposit unfit for recovery of shale oil.

# Recoverable Resources

The commercial development of an oil shale deposit depends upon many factors. The geologic setting and the physical and chemical characteristics of the resource are of primary importance. Roads, railroads, power lines, water, and available labor are among the factors to be considered in determining the viability of an oil shale operation. Oil shale lands that could be mined may be preempted by present land usage such as population centers, parks, and wildlife refuges. Development of new in-situ mining and processing technologies may allow an oil shale operation in previously restricted areas without causing damage to the surface or posing problems of air and water pollution.

The availability and price of petroleum ultimately bring about the viability of a large-scale oil shale industry. Today, few deposits can be economically mined and processed for shale oil in competition with petroleum. Nevertheless, as suppliers of petroleum diminish in future years and costs for petroleum increase, greater use of oil shale for the production of electric power, transportation fuels, petrochemicals, and other industrial products seems likely.

# Determining Grade of Oil Shale

The grade of oil shale has been determined by many different methods with the results expressed in a variety of units. The heating value of the oil shale may be determined using a calorimeter. Vales obtained by this method are reported in English or metric units, such as British thermal units (Btu) per pound of oil shale, calories per gram (cal/gm) of rock, kilocalories per kilogram (kcal/kg) of rock, megajoules per kilogram (MJ/kg) of rock, and other units. The heating value is useful for determining the quality of an oil shale that is burned directly in a power plant to produce electricity. Although the heating value of a given oil shale is a useful and fundamental property of the rock, it does not provide information on the amounts of shale oil or combustible gas that would be yielded by retorting (destructive distillation).

The grade of oil shale can be determined by measuring the yield of oil of a shale sample in a laboratory retort. This is perhaps the most common type of analysis that is currently used to evaluate an oil shale resource. The method commonly used in the United States is called the "modified Fischer assay," first developed in Germany, then adapted by the U.S. Bureau of Mines for analyzing oil shale of the Green River formation in the western United States (Stanfield and Frost, 1949). The technique was subsequently standardized as the American Society for "Testing and Materials Method D-3904-80" (1984). Some laboratories have further modified the Fischer assay method to better evaluate different types of oil shale and different methods of oil shale processing.

The standardized Fischer assay method consists of heating a 100-gram sample crushed to −8 mesh (2.38-millimeter-mesh) screen in a small aluminum retort to 500°C at a rate of 12°C per minute and held at that temperature for forty minutes. The distilled vapors of oil, gas, and water are passed through a condenser cooled with ice water into a graduated centrifuge tube. The oil and water are then separated by centrifuging. The quantities reported are the weight percentages of shale oil (and its specific gravity), water, shale residue, and "gas plus loss" by difference.

The Fischer assay method does not determine the total available energy in an oil shale. When oil shale is retorted, the organic matter decomposes into oil, gas, and a residuum of carbon char remaining in the retorted shale. The amounts of individual gases—chiefly hydrocarbons, hydrogen, and carbon dioxide—are not normally determined but are reported collectively as "gas plus loss," which is the difference of 100 weight percent minus the sum of the weights of oil, water, and spent shale. Some oil shales may have a greater energy potential than that reported by the Fischer assay method, depending on the components of the "gas plus loss."

The Fischer assay method also does not necessarily indicate the maximum amount of oil that can be produced by a given oil shale. Other retorting methods, such as the Tosco II process, are known to yield in excess of 100 percent of the yield reported by Fischer assay. In fact, special methods of retorting, such as the Hytort process, can increase oil yields of some oil shales by as much as three to four times the yield obtained by the Fischer assay method (Schora et al., 1983; Dyni et al., 1990). At best, the Fischer assay method only approximates the energy potential of an oil shale deposit.

Newer techniques for evaluating oil shale resources include the Rock-Eval and the "material balance" Fisher assay methods. Both give more complete information about the grade of oil shale but are not widely used. The modified Fischer assay, or close variations thereof, is still the major source of information for most deposits.

It would be useful to develop a simple and reliable assay method for determining the energy potential of an oil shale that would include the total heat energy and the amounts of oil, water, combustible gases including hydrogen, and char in sample residue.

# Origin of Organic Matter

Organic matter in oil shale includes the remains of algae, spores, pollen, plant cuticle and corky fragments of herbaceous and woody plants, and other cellular remains of lacustrine (i.e., sedimentary environment of a lake), marine, and land plants. These materials are composed chiefly of carbon, hydrogen, oxygen, nitrogen, and sulfur. Some organic matter retains enough biological structures so that specific types can be identified as to genus and even species. In some oil shales, the organic matter is unstructured and is best described as amorphous (bituminite). The origin of this amorphous material is not well known, but it is likely a mixture of degraded algal or bacterial remains. Small amounts of plant resins and waxes also contribute to the organic matter. Fossil shell and bone fragments composed of phosphatic and carbonate minerals, although of organic origin, are excluded from the definition of organic matter used herein and are considered to be part of the mineral matrix of the oil shale.

Most of the organic matter in oil shales is derived from various types of marine and lacustrine algae. It may also include varied admixtures of biologically higher forms of plant debris that depend on the depositional environment and geographic position. Bacterial remains can be volumetrically important in many oil shales, but they are difficult to identify.

As mentioned, most of the organic matter in oil shale is insoluble in ordinary organic solvents, whereas some is bitumen that is soluble in certain organic solvents. Solid hydrocarbons, including gilsonite, wurtzlite, grahamite, ozokerite, and albertite, are present as veins or pods in some oil shales. These hydrocarbons have somewhat varied chemical and physical characteristics, and several have been mined commercially.

# Thermal Maturity of Organic Matter

The thermal maturity of an oil shale refers to the degree to which the organic matter has been altered by geothermal heating. If the oil shale is heated to a high enough temperature, as may be the case if the oil shale is deeply buried, the organic matter may thermally decompose to form oil and gas. Under such circumstances, oil shales can be source rocks for petroleum and natural gas. The Green River oil shale, for example, is presumed to be the source of the oil in the Red Wash field in northeastern Utah. On the other hand, oil-shale deposits that have economic potential for their shale oil and gas yields are geothermally immature and have not been subjected to excessive heating. Such deposits are generally close enough to the surface to be mined by open-pit, underground mining, or by in-situ methods.

The degree of thermal maturity of an oil shale can be determined in the laboratory by several methods. One technique is to observe the changes in color of the organic matter in samples collected from varied depths in a borehole. Assuming that the organic matter is subjected to geothermal heating as a function of depth, the colors of certain types of organic matter change from lighter to darker colors. These color differences can be noted by a petrographer and measured using photometric techniques.

Geothermal maturity of organic matter in oil shale is also determined by the reflectance of vitrinite (a common constituent of coal derived from vascular land plants), if present in the rock. Vitrinite reflectance is commonly used by petroleum explorationists to determine the degree of geothermal alteration of petroleum source rocks in a sedimentary basin. A scale of vitrinite reflectance has been developed that indicates when the organic matter in a sedimentary rock has reached temperatures high enough to generate oil and gas. However, this method can pose a problem with respect to oil shale, because the reflectance of vitrinite may be depressed by the presence of lipid-rich organic matter.

Vitrinite may be difficult to recognize in oil shale because it resembles other organic material of algal origin and may not have the same reflectance response as vitrinite, thereby leading to erroneous conclusions. For this reason, it may be necessary to measure vitrinite reflectance from laterally equivalent vitrinite-bearing rocks that lock the algal material.

In areas where the rocks have been subjected to complex folding and faulting or have been intruded by igneous rocks, the geothermal maturity of the oil shale should be evaluated for proper determination of the economic potential of the deposit.

# Classification of Oil Shale

Oil shale has received many different names over the years, such as cannel coal, boghead coal, alum shale, stellarite, albertite, kerosene shale, bituminite, gas coal, algal coal, wollongite, schistes bitumineux, torbanite, and kukersite. Some of these names are still used for certain types of oil shale. Recently, however, attempts have been made to systematically classify the many different types of oil shale on the basis of the depositional environment of the deposit, the petrographic character of the organic matter, and the precursor organisms from which the organic matter was derived.

A useful classification of oil shales was developed by A. C. Hutton (1987, 1988, 1991), who pioneered the use of blue/ultraviolet fluorescent microscopy in the study of oil shale deposits in Australia. Adapting petrographic terms from coal terminology, Hutton developed a classification of oil shale based primarily on the origin of the organic matter. His classification has proved to be useful for correlating different kinds of organic matter in oil shale with the chemistry of the hydrocarbons derived from oil shale.

Hutton (1991) visualized oil shale as one of three broad groups of organic-rich sedimentary rocks: (1) humic coal and carbonaceous shale, (2) bitumen-impregnated rock, and (3) oil slate. Hutton divided oil shale into three groups based upon their environments of deposition—terrestrial, lacustrine, and marine (see figure 4.1).

Terrestrial oil shales include those composed of lipid-rich organic matter such as resin spores, waxy cuticles, and corky issue of roots, and stems of vascular plants commonly found in coal-forming swamps and bogs. Lacustrine oil shales include lipid-rich organic mater derived from algae that lived in freshwater, brackish, or saline lakes. Marine oil shales are composed of lipid-rich organic matter derived from marine algae, acritarchs (unicellular organisms of questionable origin), and marine dinoflagellates (commonly regarded as algae).

Several quantitatively important petrographic components of the organic matter in oil shale—telalginite, lamalginite, and bituminite—are adapted from coal petrogra-

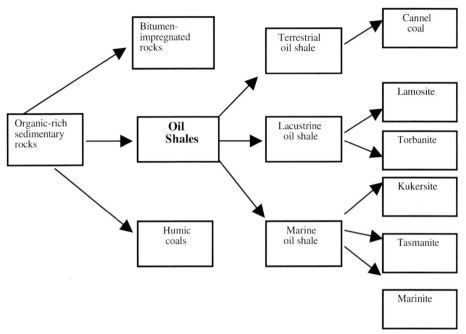

**Figure 4.1. Classification of oil shales (source: adapted from Hutton (1987).**

phy. Telalginite is organic matter derived from large colonial or thick-walled unicellular algae, typified by genera such as Botryococcus. Lamalginite includes thin-walled colonial or unicellular algae that occur as laminae with little or no recognizable biologic structures. Telalginite and lamalginite fluoresce brightly in shades of yellow under blue/ultraviolet light.

Bituminite, on the other hand, is largely amorphous, lacks recognizable biologic structures, and weakly fluoresces under blue light. It commonly occurs as an organic groundmass with fine-grained mineral matter. The material has not been fully characterized with respect to composition or origin, but it is commonly an important component of marine oil shales. Coaly materials including vitrinite and inertinite are rare to abundant components of oil shale; both are derived from humic matter of land plants and have moderate and high reflectance, respectively, under the microscope.

Within his threefold grouping of oil shales (terrestrial, lacustrine, and marine), Hutton (1991; see figure 4.1) recognized six specific oil shale types: cannel coal, Lamosite, marinite, torbanite, tasmanite, and kukersite. The most abundant and largest deposits are marinites and lamosites.

Cannel coal is brown to black oil shale composed of resins, spores, waxes, and cutinaceous and corky materials derived form terrestrial vascular plants together with varied amounts of vitrinite and inertinite. Cannel coals originate in oxygen-deficient ponds or shallow lakes in peat-forming swamps and bogs (Stach et al., 1975).

Lamosite is pale- and grayish-brown and dark gray to black oil shale in which the chief organic constituent is lamalginite derived from lacustrine planktonic algae.

Other minor components in lamosite include vitrinite, inertinite, telalginite, and bitumen. The Green River oil shale deposits in western United States and a number of the Tertiary lacustrine deposits in eastern Queensland, Australia, are lamosites.

Marinite is a gray to dark gray to black oil shale of marine origin in which the chief organic components are lamalginite and bituminite derived chiefly from marine phytoplankton. Marinite may also contain small amounts of bitumen, telalginite, and vitrinite. Marinites are deposited typically in epeiric seas (large but shallow bodies of salt water) such as on broad, shallow marine shelves or inland seas where wave action is restricted and currents are minimal. The Devonian-Mississippian oil shales of eastern United States are typical marinites. Such deposits are generally widespread, covering hundreds to thousands of square kilometers, but they are relatively thin, often less than about 100 meters.

Torbanite, tasmanite, and kukersite are related to specific kinds of algae from which the organic matter was derived; the names are based on local geographic features. Torbanite, named after Torbane Hill in Scotland, is a black oil shale whose organic matter is composed mainly of telalginite derived largely from lipid-rich Botryococcus and related algal forms found in fresh-to brackish-water lakes. It also contains small amounts of vitrinite and inertinite. The deposits are commonly small but can be extremely high grade. Tasmanite, named from oil shale deposits in Tasmania, is a brown to black oil shale. The organic matter consists of telalginite derived chiefly from unicellular tasmanitid algae of marine origin and lesser amounts of vitrinite, lamalginite, and inertinite. Kukersite, which takes its name from Kukruse Manor near the town of Kohtla-Jarve, Estonia, is a light brown marine oil shale. Its principal organic component is telalginite derived from the green alga, *Gloeocapsomorpha prisca*. The Estonian oil shale deposit in northern Estonia along the southern coast of the Gulf of Finland and its eastern extension into Russia, the Leningrad deposit, are kukersites.

# Evaluation of Oil Shale Resources

Relatively little is known about many of the world's deposits of oil shale, and much exploratory drilling and analytical work need to be done. Early attempts to determine that total size of world oil shale resources were based on few facts, and estimating the grade and quantity of many of these resources was speculative, at best. The situation today has not greatly improved, although much information has been published in the past decade or so, notably for deposits in Australia, Canada, Estonia, Israel, and the Untied States.

Evaluation of world oil-shale resources is especially difficult because of the wide variety of analytical units that are reported. The grade of a deposit is variously expressed in U.S. or Imperial gallons of shale oil per short ton (GPT) of rock, liters of shale oil per metric ton (l/t) of rock, barrels, short or metric tons of shale oil, kilocalories per kilogram (kcal/kg) of oil shale, or gigajoules (GJ) per unit weight of oil shale. To bring some uniformity into this assessment, oil shale resources in this case study are given in both metric tons of shale oil and in equivalent U.S. barrels of shale oil, and the grade of oil shale, where known, is expressed in liters of shale oil per metric ton (l/t) of rock. If the size of the resource is expressed only in volumetric units (barrels, liters, cubic metes, and so on), the density of the shale oil must be known or estimated to convert these values to metric tons. Most oil shales produce shale oil that ranges in density from about 0.85 to 0.97 by the modified Fischer assay method. In cases where the density of the shale oil is unknown, a value of 0.910 is assumed for estimating resources.

By-products may add considerable value to some oil shale deposits. Uranium, vanadium, zinc, alumina, phosphate, sodium carbonate minerals, ammonium sulfate, and sulfur are some of the potential by-products. The spent shale after retorting is used to manufacture cement, notably in Germany and China. The heat energy obtained by the combustion of the organic matter in oil shale can be used in the cement-making process. Other products that can be made form oil shale include especially carbon fibers, adsorbent carbons, carbon black, bricks, construction and decorative blocks, soil additives, fertilizers, rock wool insulating material, and glass. Most of these uses are still small or in experimental stages, but the economic potential is large.

# United States: Coal Shale Deposits

Numerous deposits of oil shale, ranging from Precambrian to Tertiary age, are present in the United States. The two most important deposits are in the Eocene Green River Formation in Colorado, Wyoming, and Utah and in the Devonian-Mississippian black shales in the eastern United States. Oil shale associated with coal deposits of Pennsylvanian age (323 to 290 million years ago) is also in the eastern United States. Other deposits are known to be in Nevada, Montana, Alaska, Kansas, and elsewhere, but these are either too small, too low grade, or have not yet been well enough explored (Russell, 1990). Because of their size and grade, most investigations have focused on the Green River and the Devonian Mississippian deposits.

## GREEN RIVER FORMATION

### Geology

Lacustrine sediments of the Green River Formation were deposited in two large lakes that occupied 65,000 square kilometers in several sedimentary-structural basins in Colorado, Wyoming, and Utah during early through middle Eocene period. The Uinta Mountain uplift and its eastward extension, the Axial Basin anticline, separate these basins. The Green River lake system was in existence for more than 10 million years during a time of a warm temperate to subtropic climate. During parts of their history, the lake basins were closed, and the waters became highly saline.

Fluctuations in the amount of inflowing stream waters caused large expansions and contractions of the lakes as evidenced by widespread intertonguing of marly lacustrine strata with beds of land-derived sandstone and siltstone. During arid times, the lakes contracted, and the waters became increasingly saline and alkaline. The lake-water content of soluble sodium carbonates and chloride increased, whereas the less soluble divalent $Ca+Mg+Fe$ carbonates were precipitated with organic-rich sediments. During the driest periods, the lake waters reached salinities sufficient to precipitate beds of nahcolite, halite, and trona. The sediment-poor waters were sufficiently saline to precipitate disseminated crystals of nahcolite, shorite, and dawsonite along with a host of other authigenic (generated where it was found or observed) carbonate and silicate minerals (Milton, 1977).

A noteworthy aspect of the mineralogy is the complete lack of authigenic sulfate minerals. Although sulfate was probably a major anion in the stream waters entering the lakes, the sulfate ion was presumably totally consumed by sulfate-reducing bacteria in the lake and sediment waters according to the following generalized oxidation-reduction reaction:

$$2CH_2O + SO_4^{-2} \rightarrow + 2HCO_3^{-1} + H_2S \text{ (hydrogen sulfide)}$$

Note that two moles of bicarbonate are formed for each mole of sulfate that is reduced. The resulting hydrogen sulfide could either react with available $Fe++$ to precipitate as ion sulfide minerals or escape from the sediments as a gas (Dyni, 1998). Other major sources of carbonate include calcium carbonate–secreting algae, hydrolysis of silicate minerals, and direct input from inflowing streams.

The warm alkaline lake waters of the Eocene Green River lakes provided excellent conditions for the abundant growth of blue-green algae (cyanobacteria) that are thought to be the major precursor of the organic matter in the oil shale. During times of freshening waters, the lakes hosted a variety of fishes, rays, bivalves, gastropods, ostracodes, and other aquatic fauna. Areas peripheral to the lakes supported a large and varied assemblage of land plants, insects, amphibians, turtles, lizards, snakes, crocodiles, birds, and numerous mammalian animals (McKenna, 1960; MacGinitie, 1969; and Grande, 1984).

### Historical Developments

The occurrence of oil shale in the Green river Formation in Colorado, Utah, and Wyoming has been known for many years. During the early 1900s, it was clearly established that the Green River deposits were a major resource of shale oil (Woodruff and Day, 1914; Winchester, 1916; Gavin, 1924). During this early period, the Green River and other deposits were investigated, including oil shale of the marine Phosphoria Formation of Permian age in Montana (Bowen, 1917; Condit, 1919) and oil shale in Tertiary lake beds near Elko, Nevada (Winchester, 1923).

In 1967, the U.S. Department of Interior began an extensive program to investigate the commercialization of the Green River oil shale deposits. The dramatic increases in petroleum prices resulting from the OPEC oil embargo of 1973–1974 triggered another resurgence of oil shale activities during the 1970s and into the early 1980s. In 1974 several parcels of public oil shale lands in Colorado, Utah, and Wyoming were put up for competitive bid under the Federal Prototype Oil Shale Leasing

Program. Two tracts were leased in Colorado (C-a and C-b) and two in Utah (U-a and U-b) to oil companies.

Large underground mining facilities, including vertical shafts, room-and-pillar entries, and modified in situ retorts, were constructed on Tracts C-a and C-b, but little or no shale oil was produced. During this time, Unocal Oil Company was developing its oil shale facilities on privately owned land on the south side of the Piceance Creek Basin. The facilities included a room-and-pillar mine with a surface entry, a 10,000-barrel/day (1,460-ton/day) retort, and an upgrading plant. A few miles north of the Unocal property, Exxon Corporation opened a room-and-pillar mine with a surface entry, haulage roads, waste-rock dumpsite, and a water-storage reservoir and dam.

In 1977–1978 the U.S. Bureau of Mines opened an experimental mine that included a 723-meter-deep shaft with several room-and-pillar entries in the northern part of the Piceance Creek Basin to conduct research on the deeper deposits of oil shale, which are commingled with nahcolite and dawsonite. The site was closed in the later 1980s.

About $80 million were spent on the U-a/U-b tracts in Utah by three energy companies to sink a 313-meter-deep vertical shaft and inclined haulage way to a high-grade zone of oil shale and to open several small entries. Other facilities included a mine services building, water- and sewage-treatment plants, and a water-retention dam.

The Seep Ridge project, sited south of the U-a/U-b tracts and funded by Geokinetics, Inc. and the U.S. Department of Energy, produced shale oil by a shallow in situ retorting method. Several thousand barrels of shale oil were produced.

The Unocal oil shale plant was the last major project to produce shale oil from the Green River Formation. Plant construction began in 1980, and capital investment for constructing the mine, retort, upgrading plant, and other facilities was $650 million. Unocal produced 657,000 tons (about 4.4 million barrels) of shale oil, which were shipped to Chicago for refining into transportation fuels and other products under a program partly subsidized by the U.S. government. The average rate of production in the last months of operation was about 875 tons (about 5,900 barrels) of shale oil per day; the facility was closed in 1991.

In the past few years, Shell Oil Company began an experimental field project to recover shale oil by a proprietary in situ technique. Some details about the project have been publicly announced, and the results to date (2006) appear to favor continued research.

### Shale Oil Resources

As the Green River oil shale deposits in Colorado became better known, estimates of the resource increased from about 20 billion barrels in1916, to 900 billion barrels in 1961, and to 1.0 trillion barrels (−147 billion tons) in 1989 (Winchester, 1916; Donnell, 1961; Pitman et al., 1989).

The Green River oil shale resources in Utah and Wyoming are not as well known as those in Colorado. Trudell et al. (1983) calculated the measured and estimated resources of shale oil in an area of about 5,200 square kilometers in eastern Uinta Basin, Utah, to be 214 billion barrels (31 billion tons) of which about one-third is in the rich Mahogany oil shale zone. Culbertson et al. (1980) estimated the oil shale resources in the Green River Formation in the Green River Basin in southwest Wyoming to be 244 billion barrels (35 billon tons) of shale oil.

Additional resources are also in the Washakie Basin east of the Green River Basin in southwest Wyoming. Trudell et al. (1973) reported that several members of the Green River Formation on Kinney Rim on the west side of the Washakie Basin contain sequences of low to moderate grades of oil shale in three core holes. Two sequences of oil shale in the Laney Member, eleven and forty-two meters thick, average 63 l/t and represent as much as 8.7 million tons of in-situ shale oil per square kilometer.

*Other Mineral Resources*

In addition to fossil energy, the Green River oil shale deposits in Colorado contain valuable resources of sodium carbonate minerals including nahcolite ($NaHCO_3$) and dawsonite [$NaAl(OH)_2CO_3$]. Both minerals are commingled with high-grade oil shale in the deep northern part of the basin. Dyni (1974) estimated the total nahcolite resource at 29 billion tons. Beard et al. (1974) estimated nearly the same amount of nahcolite and 17 billion tons of dawsonite. Both minerals have value for soda as ($Na_2CO_3$) and dawsonite also has potential value for its alumina ($Al_2O_3$) content. The latter mineral is most likely to be recovered as a by-product of an oil shale operation. One company is solution-mining nahcolite for the manufacture of sodium bicarbonate in the northern part of the Piceance Creek Basin at depths of about 600 meters (Day, 1998). Another company stopped solution-mining nahcolite in 2004, but now processes soda ash obtained from the Wyoming trona (hydrated sodium bicarbonate carbonate) deposits to manufacture sodium bicarbonate.

The Wilkins Peak Member of the Green River Formation in the Green River Basin in southwestern Wyoming contains not only oils but also the world's largest known source of natural sodium carbonate as trona ($Na_2CO_3 \cdot NaHCO_3 \cdot 2H_2O$). The trona resource is estimated at more than 115 billion tons in twenty-two beds ranging from 1.2 to 9.8 meters in thickness (Wiig et al., 1995). In 1997, trona production from five mines was 16.5 million tons (Harris, 1997). Trona is refined into soda ash ($Na_2CO_3$) used in the manufacture of bottle and flat glass, baking soda, soap and detergents, waste treatment chemicals, and many other industrial chemicals. One ton of soda ash is obtained from about two tons of trona ore. Wyoming trona supplies about 90 percent of U.S. soda ash needs; in addition, about one-third of the total Wyoming soda ash produced is exported.

In the deeper part of the Piceance Creek Basin, the Green River oil shale contains a potential resource of natural gas, but its economic recovery is questionable (Cole and Daub, 1991). Natural gas is also present in the Green River oil shale deposits in southwest Wyoming, and probably in the oil shale in Utah, but in unknown quantities.

# EASTERN DEVONIAN-MISSISSIPPIAN OIL SHALE

## Depositional Environment

Black organic-rich marine shale and associated sediments of Late Devonian and Early Mississippian age underlie about 725,000 square kilometers in the eastern United States. These shales have been exploited for many years as a resource of natural gas, but have also been considered as a potential low-grade resource of shale oil and uranium (Roen and Kepferle, 1993; Conant and Swanson, 1961).

Over the years, geologists have applied many local names to these shales and associated rocks, including the Chattanooga, New Albany, Ohio, Sunbury, Antrim, and others. A group of papers detailing the stratigraphy, structure, and gas potential of these rocks in eastern United States have been published by the U.S. Geological Survey (Roen and Kepferle, 1993).

The black shales were deposited during Late Devonian and Early Mississippian times in a large epeiric sea that covered much of middle and eastern United States east of the Mississippi River. The area includes the broad, shallow, Interior Platform on the west that grade eastward into the Appalachian Basin. The depth to the base of the Devonian-Mississippian black shales ranges from surface exposures on the Interior Platform to more than 2,700 meters along the depositional axis of the Appalachian Basin (De Witt et al., 1993).

The Late Devonian sea (Paleozoic era; about 416–2.5 million years ago) was relatively shallow with minimal current and wave action, much like the environment in which the Alum Shale of Sweden was deposited in Europe. A large part of the organic matter in the black shale is amorphous bituminite, although a few structured fossil organisms such as Tasmanites, Botryococcus, Foerstia, and others have been recognized. Conodonts (extinct primitive fishlike chordates with conelike teeth) and linguloid (small, oval, and inarticulate) brachiopods are sparingly distributed through some beds. Although much of the organic matter is amorphous and of uncertain origin, it is generally believed that much of it was derived form planktonic algae.

In the distal (furthest away) parts of the Devonian sea, the organic matter accumulated very slowly along with very fine-grained clayey sediments in poorly oxygenated waters free of burrowing organisms. Conant and Swanson (1961) estimated that thirty centimeters of the upper part of the Chattanooga Shale deposited on the Interior Platform in Tennessee could represent as much as 150,000 years of sedimentation.

The black shales thicken eastward into the Appalachian Basin owning to increasing amounts of clastic sediments that were shed into the Devonian sea from the Appalachian highland lying to the east of the basin. Pyrite and marcasite are abundant authigenic materials, but carbonate minerals are only a minor fraction of the mineral matter.

## Resources

The oil shale resource is in that part of the Interior Platform where the black shales are the richest and closest to the surface. Although long known to produce oil upon retorting, the organic matter in the Devonian-Mississippian black shale yields only about half as much as the organic matter of the Green River oil shale, which is thought to be attributable to differences in the type of organic matter for (or type of organic carbon in) each of the oil shales. The Devonian-Mississippian oil shale has a higher ratio of aromatic to aliphatic organic carbon than Green River oil shale, and is shown by material balance Fischer assays to yield much less shale oil and a higher percentage of carbon residue (Miknis, 1990).

Hydroretorting Devonian-Mississippian oil shale can increase the oil yield by more than 200 percent of the value determined by Fischer assay. In contrast, the conversion of organic matter to oil by Hydroretorting is much less for Green River oil shale, about 130 to 140 percent of the Fischer assay value. Other marine oil shales also respond favorably to hydroretorting, with yields as much as, or more than, 300 percent of the Fischer assay (Dyni et al., 1990).

Matthews et al. (1980) evaluated the Devonian-Mississippian oil shales in areas of the Interior Platform where the shales are rich enough in organic matter and close enough to the surface to be mineable by open pit. Results or investigations in Alabama, Illinois, Indiana, Kentucky, Ohio, Michigan, eastern Missouri, Tennessee, and West Virginia indicated the 98 percent of the near-surface mineable resources are in Kentucky, Ohio, Indiana, and Tennessee (Matthews, 1983).

The criteria for the evaluation of the Devonian-Mississippian oil shale resource used by Matthews et al. (1980) were:

1. Organic carbon content: $\geq 10$ weight percent
2. Overburden: $\leq 200$ m
3. Stripping ratio: $\leq 2.5{:}1$
4. Thickness of shale bed: $\geq 3$ m
5. Open-pit mining and hydroretorting

On the basis of these criteria, the total Devonian-Mississippian shale oil resources were estimated to be 423 billion barrels (61 billion tons).

# Chemical Sedimentary Rocks

Chemical and organic sedimentary rocks are the other main group of sediments besides clastic sediments. They are formed by weathered material in solution precipitating from water or as biochemical rocks made of dead marine organisms. Usually special conditions are required for these rocks to form, such as high temperature, high evaporation, and high organic activity. Some chemical sediment is deposited directly from the water in which the material is dissolved; for example, solution upon evaporation of seawater. Such deposits are generally referred to as inorganic chemical sediments. Chemical sediments that have been deposited by or with the assistance of plants or animals are said to be organic or biochemical sediments. Accumulated carbon-rich plant material may form coal. Deposits made mostly of animal shells may form limestone, chert, or coquina.

## CHEMICAL SEDIMENTARY ROCKS

Sedimentary rocks formed from sediments created by inorganic processes are discussed below.

**Limestone**—calcite ($CaCO_3$) is precipitated by organisms usually to form a shell or other skeletal structure. Accumulation of these skeletal remains results in the most common type of chemical sediment, limestone. Limestone may form by inorganic precipitation as well as by organic activity.

**Dolomite**—consists of carbonate mineral (known as magnesium limestone $[CaMg(CO_3)_2]$), and occurs in more or less the same settings as limestone. Dolomite is formed when some of the calcium in limestone is replaced by magnesium.

**Evaporites**—these are sedimentary rocks (true chemical sediments) that are derived from minerals precipitated from seawater. Thus, they consist mostly of halite (salt [$NaCl$]) and gypsum ($CaSO_4$) by chemical precipitation—high evaporation rates cause concentration of solids to increase due to water loss by evaporation.

## BIOCHEMICAL SEDIMENTARY ROCKS

These rocks consist of sediments formed from the remains and/or secretions of organisms including **fossiliferous limestone**, **coquina** (type of limestone composed of shells and coarse shell fragments), **chalk** (a porous, fine-textured variety of limestone composed of calcareous shells), **lignite** (brown coal), and **bituminous** (soft) **coal** (see Case Study 4.2).

~

### Case Study 4.2—Coal[2]

America has more coal than any other fossil fuel resource. The United States also has more coal reserves than any other single country in the world. In fact, one-fourth of all the known coal in the world is in the United States. The United States has more coal that can be mined than the rest of the world has oil that can be pumped from the ground.

Currently, coal is mined in 26 of the 50 states.

Coal is used primarily in the United States to generate electricity. In fact, it is burned in power plants to produce more than half of the electricity we use. A stove uses about half a ton of coal a year. A water heater uses about two tons of coal a year. Add a refrigerator, that's another half-ton a year. Even though you many never see coal, you use several tons of it every year!

The material that formed fossil fuels varied greatly over time as each layer was buried. As a result of these variations and the length of time the coal was forming, several types of coal were created. Depending upon its composition, each type of coal burns differently and releases different types of emissions.

The four types (or ``ranks'') of coal mined today are: anthracite, bituminous, sub-bituminous and lignite.

- **Lignite**: The largest portion of the world's coal reserves is made up of lignite, a soft, brownish-black coal that forms the lowest level of the coal family. You can even see the texture of the original wood in some pieces of lignite that is found primarily west of the Mississippi River in the United States.
- **Sub-bituminous**: Next up the scale is sub-bituminous coal, a dull black coal. It gives off a little more energy (heat) than lignite when it burns. It is mined mostly in Montana, Wyoming, and a few other western states.
- **Bituminous**: Still more energy is packed into bituminous coal, sometimes called ''soft coal.'' In the United States, it is found primarily east of the Mississippi River in midwestern states like Ohio and Illinois and in the Appalachian mountain range from Kentucky to Pennsylvania.
- **Anthracite**: This is the hardest coal and gives off a great amount of heat when it burns. Unfortunately, in the United States, as elsewhere in the world, there is little anthracite coal to be mined. The U.S. reserves of anthracite are located primarily in Pennsylvania.

# Physical Characteristics of Sedimentary Rocks

Sedimentary rocks possess definite physical characteristics and display certain features that make them readily distinguishable from igneous or metamorphic rocks. Some of the most important sedimentary characteristics are:

- **Stratification**—probably the most characteristic feature of sedimentary rocks is their tendency to occur in strata, or beds. These strata are formed as geological agents such as wind, water, or ice gradually deposit sediment
- **Cross-bedding**—sets of beds that are inclined relative to one another. The beds are inclined in the direction that the wind or water was moving at the time of deposition. Boundaries between sets of cross beds usually represent an erosional surface. Very common in beach deposits, sand dunes, and river-deposited sediment.
- **Texture**—the size, shape, and arrangements of materials, which are derived by the nature of the source materials, wind and water currents present, the distance materials are transported or time in the transportation process, biological activity, and exposure to various chemical environments.
- **Graded bedding**—bedding showing a decrease in grain size from the bottom of the bed to the top of the bed (fine sediment on top and coarse at bottom). In a stream, as current velocity wanes, first the larger or denser particles are deposited, followed by smaller particles.
- **Ripple marks**—characteristic of shallow-water deposition. Caused by small waves or winds, such as those that cause ripples of sand on the surface of a beach or bottom of a stream. Ripples of this type have also been preserved in certain sedimentary rocks and may provide information about the conditions of deposition when the sediment was originally deposited.

- **Mud cracks**—resulting from the drying out of wet sediment on the bottom of dried-up lakes, ponds, or stream beds. These many-sided (polygonal) shapes give a honeycomb appearance on the surface. If preserved in sedimentary rocks, such shapes suggest that the rock was subjected to alternate periods of flooding and drying.
- **Concretions**—spherical or flattened masses of rock enclosed in some shales or limestones that are generally harder than the rock enclosing them. Because concretions are usually harder than the enclosing rock, they are often left behind after the surrounding rock has been eroded away.
- **Fossils**—remains or evidence of ancient organisms that have been preserved in the earth's crust. Because life has evolved, fossils give clues to the relative age of the sediment; they can be important indicators of past climates.
- **Color**—such as that produced by hematites (iron oxides), giving a pink or red color in such areas as the Grand Canyon and Painted Desert.

# Sedimentary Rock Facies

A sedimentary facies is a group of characteristics that describe an accumulation of deposits that have distinctive characteristics and grade laterally into other sedimentary deposits as a result of changing environments and original deposits.

# Chapter Review Questions

4.1   Eroded rock fragments: _____.
4.2   Sedimentary rocks are made up of pieces of preexisting _____.
4.3   _____ is the process of chemical and physical change that turns sediment into rock.
4.4   _____ are consolidated gravel deposits.
4.5   Chemical sediments that have been deposited by or with the assistance of plants or animals are said to be organic or _____ _____.
4.6   Organisms' shells or skeletal structures can form _____.
4.7   Derived from mineral precipitated from seawater: _____.
4.8   Soft, brownish-black coal that forms the lowest level of the coal family: _____.
4.9   Sets of rock beds that are inclined relative to one another: _____.
4.10  Spherical or flattened masses of rock enclosed in some shales or limestone: _____.

# References and Recommended Reading

ASTM. 1966. Designated D 388-66—Specifications for classification of coals by rank. In *Annual Book of American Society for Testing and Materials Standards*, West Conshohocken, PA: ASTM, pp. 66–71.

ASTM. 1984. Designation D 3904-80—Standard test method for oil from oil shale. In *Annual Book of American Society for Testing and Materials Standards*. West Conshohocken, PA: ASTM, pp. 513–25.

Beard, T. M., Tait, D. B., and Smith, J. W. 1974. *Nahcolite and Dawsonite Resources in the Green River Formation, Piceance Resources of the Piceance Creek Basin, 25th Field Conference*. Denver: Rocky Mountain Association of Geologists, pp. 101–9.

Bowen, F. F. 1917. *Phosphatic Oil Shales Near Dell and Dillon, Beaverhead Country, Montana*: U.S. Geological Survey Bulletin 661, pp. 315–20.

Cole, R. D., and Daub, G. T. 1991. Methane occurrences and potential resources in the lower Parachute Creek Meander of Green River Formation, Piceance Creek Basin, Colorado. In 24th Oil Shale Symposium Proceedings. *Colorado School of Mines Quarterly* 83, no. 4, pp. 1–7.

Conant, L. C., and Swanson, V. E. 1961. *Chattanooga Shale and Related Rocks of Central Tennessee and Nearby Areas*: U.S Geological Survey Professional Paper 357.

Condit, D. D. 1919. *Oil Shale to Western Montana, Southeastern Idaho, and Adjacent Parts of Wyoming and Utah*. U.S. Geological Survey Bulletin 711, pp. 15–40.

Cross, T. A., and Homewood, P. W. 1997. Amanz Gressly's role in founding modern stratigraphy. *Geological Society of America Bulletin* 109, no. 12, pp. 1617–30.

Culbertson, W. C., Smith, J. W., and Trudell, L. G. 1980. *Oil Shale Resources and Geology of the Green River Formation in the Green River Basin, Wyoming*. U.S Department of Energy Laramie Energy Technology Center LETC/RI-80/6.

Day, R. L. 1998. Solution mining of Colorado nahcolite. In *Proceedings of the First International Soda Ash Conference, Rocks Springs, Wyoming, June 10–12, 1997*. Wyoming State Geological Survey Public Information Circular 40.

De Witt, W. Jr., Roen, J. B., and Wallace, L. G. 1993. Stratigraphy of Devonian black shales and associated rocks in the Appalachian Basin. In *Petroleum Geology of the Devonian and Mississippian Black Shale of Eastern North America*. U.S. Geological Survey Bulletin 1909, Chapter B, pp. B1–B57.

Donnell, J. R. 1961. *Tertiary Geology and Oil-Shale Resources of the Piceance Creek Basin between the Colorado and White Rivers, Northwestern Colorado*. U.S. Geological Survey Bulletin 1082-L, pp. 835–91.

Dyni, J. R. 1974. Stratigraphy and nahcolite resources of the saline facies of the Green River Formation in northwest Colorado. In *Guidebook to the Energy Resources of the Piceance Creek Basin, 25th Field Conference*. Denver, CO: Rocky Mountain Association of Geologists, pp. 111–22.

Dyni, J. R., Anders, D. E., and Rex, R. C. Jr. 1990. Comparison of hydro-retorting, Fischer assay, and Rock-Eval analyses of some world oil shales. In *Proceedings 1989 Eastern Oil Shale Symposium*. Lexington: University of Kentucky Institute of Mining and Minerals Research, pp. 270–86.

Dyni, J. R. 1998. Prospecting for Green River-type sodium carbonate deposits. In *Proceedings of the First International Soda Ash Conference*, vol. II. Wyoming State Geological Survey Information Circular 40, pp. 37–47.

Fichter, L. 2000. A basic sedimentary rock classification. Accessed 05/31/08 at http://csmres .jmu.edu/geollab/fichter/SedRx/sedclass.html.

Gavin, M. J. 1924. *Oil shale: An Historical, Technical, and Economic Study*: U.S. Bureau of Mines Bulletin 210.

Grande, L. 1984. *Paleontology of the Green River Formation with a Review of the Fish Fauna*, 2nd ed. Geological Survey of Wyoming Bulletin 63.

Harris, R. E. 1997. Fifty years of Wyoming trona mining. In *Prospect to Pipeline*, 48th Guidebook. Casper: Wyoming Geological Association, pp. 177–82.

Hutton, A. C. 1987. Petrographic classification of oil shales. *International Journal of Coal Geology* 8, pp. 203–31.

Hutton, A. C. 1988. Organic petrography of oil shales: U.S. Geological Survey short course, January 25–29, Denver, CO [unpublished].

Hutton, A. C. 1991. Classification, organic petrography and geochemistry of oil shale. In Proceedings 1990 Eastern Oil Shale Symposium: Lexington, University of Kentucky Institute for Mining and Minerals research, pp. 16–172.

MacGinitie, H. D. 1969. *The Eocene Green River Flora of Northwestern Colorado and Northeastern Utah*. Berkeley: University of California Press.

Matthews, R. D. 1983. The Devonian-Mississippian oil shale resource of the Untied States. In Gary, H. H., ed., *Sixteenth Oil Shale Symposium Proceedings*. Golden: Colorado School of Mines Press, pp. 14–25.

Matthews, R. D., Janka, J. C., and Dennison, J. M. 1980. *Devonian Oil Shale of the Eastern United States, a Major American Energy Resource* [preprint]: Evansville, IN, American Association of Petroleum Geologists Meeting, Oct. 1–3, 1980.

McKenna, M. C. 1960. *Fossil Mammalia from the Early Wasatchian Four Mile Fauna, Eocene of Northwest Colorado*. Berkeley: University of California Press.

Miknis, F. P. 1990. Conversion characteristics of selected foreign and domestic oil shales. In *Twenty-third Oil Shale Symposium Proceedings*. Golden: Colorado School of Mines Press, pp. 100–109.

Milton, C. 1977. Mineralogy of the Green River Formation. *The Mineralogy Record* 8, pp. 368–79.

Pitman, J. K., Pierce, F. W., and Grundy, W. D. 1989. *Thickness, Oil-Yield, and Kriged Resource Estimates for the Eocene Green River Formation, Piceance Creek Basin, Colorado*. U.S. Geological Survey Oil and Gas Investigations Cart OC-123.

Reading, H. G. (Ed.). 1996. *Sedimentary Environments and Facies*. New York: Blackwell Scientific Publications.

Roen, J. B., and Kepferle, R. C., eds. 1993. *Petroleum Geology of the Devonian and Mississippian Black Shale of Eastern North America*. U.S. Geological Survey Bulletin 1909, chapters A through N.

Russell, P. L. 1990. *Oil Shales of the World: Their Origin, Occurrence and Exploitation*. New York: Pergamon Press.

Schora, F. C., Janka, J. C., Lynch, P. A., and Feldkirchner, H. 1983. Progress in the commercialization of the Hytort Process. In Proceedings 1982 Eastern Oils Shale Symposium: Lexington, University of Kentucky, Institute for Mining and Minerals Research, pp. 183–90.

Stach, E., Taylor, G. H., Machowsky, M.-Th., Chandra, D., Teichmuller, M., and Teichmuller, R. 1975. *Stach's Textbook of Coal Petrology*. Berlin: Gebruder Borntradger.

Stanfield, K. E., and Frost, I. C., 1949. *Method of Assaying Oil Shale by a Modified Fischer Retort*. U.S. Bureau of Mines Report of Investigations 4477.

Trudell, L. G., Roehler, H. W., and Smith, J. W. 1973. *Geology of Eocene Rocks and Oil Yields of Green River Oil Shales on Part of Kinney Rim, Washakie Basin, Wyoming*. U.S. Bureau of Mines Report of Investigations 7775.

Trudell, L. G., Smith, J. W., Beard, T. N., and Mason, G. M. 1983. *Primary Oil-Shale Resources of the Green River Energy Laramie Energy Technology Center*. DOE/LC/RI-82-4.

USGS. 2004. Sedimentary rocks. Accessed 06/02/08 at http://geomaps.wer.usgs.gov/par/rxmin/rock2.html.

USGS. 2006. The making of sedimentary rocks. Accessed 06/03/08 at http://education.usgs.gov/Schoolyard/rocks sedimentary.html.

Wiig, S. V., Grundy, W. D., and Dyni, J. R. 1995. Trona resources in the Green River Formation, southwest Wyoming: U.S. Geological Survey Open-File Report 95-476.

Winchester, D. E. 1916. *Oil Shale in Northwestern Colorado and Adjacent Areas.* U.S. Geological Survey Bulletin 641-F, pp. 139–98.

Winchester, D. E. 1923. *Oil Shale of the Rocky Mountain Region.* U.S. Geological Survey Bulletin 729.

Woodruff, E. G., and Day, D.T. 1914. Oil shales of northwestern Colorado and northeastern Utah: U.S. Geological Survey Bulletin 581, p. 1.

# Notes

1. The material presented in this section is from Jon R. Dyni, *Geology and Resources of Some World Oil-Shale Deposits* (Denver, CO: USGS Central Region, 2005).

2. The information in the case study is from U.S. Department of Energy, "Coal: Our Most Abundant Fuel," accessed 06/09/08 at http://fossil.energy.gov/education/energylessons/coal/gen_coal. html.

# CHAPTER 5

# Metamorphism, Metamorphic Rocks, and Deformation

In geology, metamorphism (Gr. *meta* = "change," *morph* = "form," so metamorphism means to change form) is the process of change (in mineral assemblage and texture) that rocks within the earth undergo when exposed to increasing temperatures and pressures at which their mineral components are no longer stable. Metamorphism may be *local*—contact metamorphism is due to igneous intrusion—or *regional*—as takes place in mountain building, when slate, schist, and gneiss are formed. Contrary to the popular view that metamorphism can't occur unless tremendous heat is generated, metamorphism may take place in a solid state, without melting.

Recall that in our earlier discussion of sedimentary rocks it was pointed out that sedimentary rocks also go through a process of changing form known as **diagenesis**. In geology, however, we restrict these sedimentary processes to those that occur at temperatures below 200°C and pressures below about 300 mega pascal (MPa); this is equivalent to approximately 3,000 atmospheres of pressure. On the other hand, metamorphism occurs at temperatures and pressure higher than 200°C and 300 MPa. The upper limit of metamorphism occurs at the pressure and temperature of wet partial melting of the affected rock. Keep in mind, however, that once melting begins the process changes to an igneous process.

## Metamorphism: Source of Heat and Pressure

Metamorphism occurs because some minerals are stable only under certain conditions of pressure and temperature. The heat involved with metamorphic processes is derived from uranium and thorium and other elements with lead and radiation added. Potassium-40 to calcium-40 or argon plus radiation provides another internal source of heat. In regards to pressure, recall that air pressure at sea level is measured at fourteen pounds per square inch (1 atmosphere [atm] or 1 bar = 100,000 pascals). In comparison, the pressure beneath thirty-three feet of water is equal to 1 atmosphere (1 bar). It is interesting to note that this same level of pressure is measured beneath ten feet of rock. Pressure measured in the deepest part of ocean is equivalent to 1,000 bars. The pressure measured under one mile of rock is 500 bars (1,000 bars or 1 kilobar (kb) is the measured pressure beneath two miles of rock). The point is that rocks can be subjected to higher temperatures and pressure as they become buried deeper in the earth. Such burial usually takes place as a result of tectonic processes such as continental collisions or subduction (discussed in detail later). These same processes (especially tectonic uplift and erosion) eventually are involved with projecting metamorphic rocks to the surface.

# Types of Metamorphism

As mentioned, metamorphism consists of two types: contact metamorphism and regional metamorphism.

- **Contact metamorphism**—occurs around or adjacent to igneous intrusions under low pressure but high temperatures in local, shallow areas of 0–6 kilometers. Because only a small area surrounding the intrusion is heated by the magma, metamorphism is restricted to a zone surrounding the intrusion.
- **Regional metamorphism**—occurs over large areas that were subjected to high pressure, causing deformation. Strongly foliated metamorphic rocks such as schists, slates, and gneisses are formed.

## Did You Know?

Limestone is formed by the mineral calcite. Calcite is very stable over a wide range of temperatures and pressures. Consequently, when metamorphism of limestone occurs, the original calcite crystals grow larger. The resultant rock produced? Marble.

# Major Metamorphic Rock Types

The three major metamorphic rock types include slate, schist, and gneiss.

- **Slate**—is fine-grained chlorite and clay minerals; it is generally a foliated (banded), homogenous, metamorphic rock derived from an original sedimentary shale form through low-grade regional metamorphism. Figure 5.1 shows the response of shale rock to increasing metamorphic processes.
- **Schist**—is a crystalline rock formed primarily from basalt, an igneous rock; shale, a sedimentary rock; or slate, a metamorphic rock. Tremendous heat and pressure caused these rocks to be transformed into schist rock. Schist rock has the tendency to split into irregular planar layers (schistosity). Most schist is composed largely of platy minerals such as muscovite, chlorite, talc, biotite, and graphite. Figure 5.2 shows schistosity of metamorphic schist rock.
- **Gneiss**—(pronounced like your relative) is a metamorphic rock—metamorphosed primarily from granite and diorite—characterized by banding caused by segregation of different types of rocks, typically light and dark silicates. In the formation of gneiss, metamorphism continues and the sheet silicates become unstable and dark-colored minerals like hornblende and pyroxene start to grow. As mentioned, these dark-colored minerals tend to become segregated in distinct bands through the rock, giving the rock a gneissic banding. The name gneissic actually is more suitable for its texture. Again, gneissic texture refers to the segregation of light and dark minerals. This light and dark segregation is shown in figure 5.3.

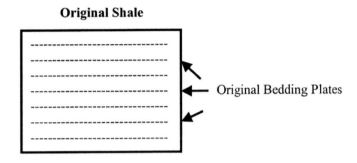

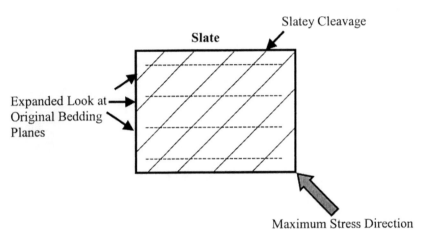

**Figure 5.1.** Metamorphism of clay and quartz minerals (shale) where maximum stress is applied at an angle to the original bedding planes, so that the slatey cleavage has developed at an angle to the original bedding.

## Metamorphic Facies

As mentioned earlier in our discussion of sedimentary facies, a **facies** is a body of rock with specified characteristics. Metamorphic facies is similar to the concept of sedimentary facies, in that a sedimentary facies is also a set of environmental conditions present during deposition. Keep in mind, however, that in general, metamorphic rocks do not undergo significant changes in chemical composition during metamorphism. In metamorphism—changes in rock assemblages—it is all about pressure and temperature, and variations from location to location, from rock type to rock type, and assemblage to assemblage. Consequently, the mineral assemblages that are observed are an indication of the temperature and pressure environment that they were subjected to. Moreover, it is important to keep in mind that though we are speaking of changes in rocks because of metamorphism, not all changes in rocks are due to metamorphism; changes also occur due to surface agents such as weathering (discussed later) and sedimentary processes such as diagenesis (discussed earlier).

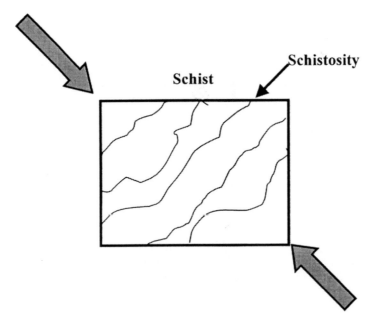

**Figure 5.2.   The irregular planar foliation (schistosity) of metamorphic schist rock.**

# Deformation of Rock

Although rock beds, especially sedimentary rocks, are generally deposited in almost horizontal layers, they are generally found to be tilted or distorted if followed over any considerable distance. Such structures are the result of forces within the earth that continually bend them, twist them, or fracture them. When rocks bend, twist, or fracture we say they *deform*. When rocks deform (change in size, shape, or volume) they are said to *strain*. The intensity of the changes is proportional to the intensity of deformation and depth of burial. The forces that cause deformation of rock are referred to as stresses. The most intense zones of rock deformation are associated with mountain-building processes.

To better understand rock deformation we must first understand the forces of stress.

- **Stress**—a force applied over an area (force/unit area).
- **Uniform stress**—a stress wherein the forces act equally from all directions (also called pressure). The pressure due to the weight of overlying rocks within the earth is uniform stress but is usually referred to as *confining stress*.
- **Differential stress**—a stress when the stress is not equal from all directions. Three kinds of differential stress occur (see figure 5.4).
  1. Tensional or extensional stress—stretches rock.

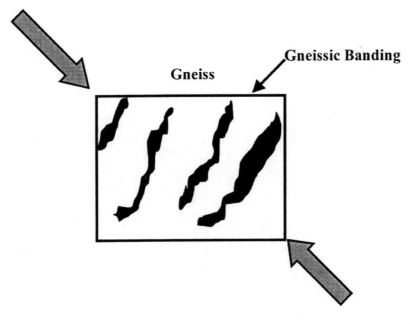

**Figure 5.3.** In gneiss the dark-colored minerals tend to form elongated crystals; however, they still have a preferred orientation with their long directions perpendicular to the maximum differential stress.

2. Compressional stress—squeezes rock.
3. Shear stress—results in slippage and translation.

## TYPES OF DEFORMATION

Increasing stress causes rock to deform in three successive stages (see figure 5.5).

- **elastic deformation**—strain is reversible
- **ductile deformation**—strain is irreversible
- **fracture**—irreversible strain wherein the material breaks

## Did You Know?

Brittle materials have a small or large region of elastic behavior but only a small region of ductile behavior before they fracture. On the other hand, ductile materials have a small region of elastic behavior and a large region of ductile behavior before they fracture.

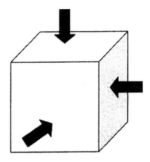

**Confining Stress** (Stress equal from all directions)

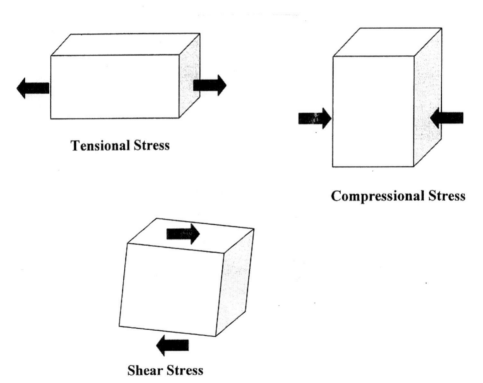

**Tensional Stress**

**Compressional Stress**

**Shear Stress**

Figure 5.4.    Kinds of differential stress.

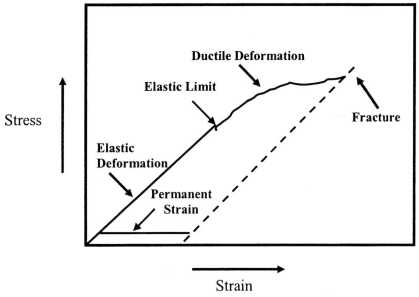

**Figure 5.5.  Stages of rock deformation.**

How a particular material behaves depends on several factors, including:

- **Temperature**—when temperature is high, materials are ductile. At low temperature, materials are brittle.
- **Confining pressure**—high surrounding pressure tends to hinder fracture. At low confining pressure, material fractures sooner.
- **Strain rate**—high strain rates cause materials to fracture. Low strain rates favor ductile behavior.
- **Composition**—deals with the chemical bond types that hold rocks together. Quartz, olivine, and feldspars are very brittle, while others such as micas, clay, and calcite minerals are more ductile. The presence of water is another compositional factor. Water weakens chemical bonds and allows rock to act in a ductile manner; dry rocks, however, tend to behave in a brittle manner.

## Did You Know?

Geologists study rock deformation for many reasons. For example, it is important to know how beds and rocks are deformed in order to determine location of coal seams, water aquifers, etc. Also, the location of ore deposits and petroleum traps is also of economic importance.

## DEFORMATION IN PROGRESS

Deformation of rocks is an ongoing process. However, to the casual observer this ongoing deformation process usually is not evident because it is slow and gradual. In many cases it is so slow that it can only be observed and documented by using sensitive measuring instruments over periods of months, years, and decades, or even longer. Again, unless deformation is abrupt along fault lines (earthquake) caused by the fracture of rocks on a time scale of seconds or minutes, we are unlikely to notice deformation in progress.

## GEOLOGICAL EVIDENCE OF FORMER DEFORMATION

When we observe crustal rock formations, evidence of deformation that has occurred in the past is very evident. For example, lava flows and sedimentary strata generally follow the law of original horizontality (sediments pile up sandwich-fashion and lava simply piles up). Thus, when we observe strata that is no longer horizontal but instead is inclined, it is clear that deformation has occurred some time in the past. Geologists are able to uniquely define and communicate to others the orientation of planar features by using two special terms—strike and dip.

Dip of a bed is a measure of its slope or tilt in relation to the horizontal. For an inclined plane, the **strike** is the compass direction of any horizontal line on the plane. Again, stated in relation to strike, the dip is the angle between a horizontal plane and the inclined plane, measured perpendicular to the direction of strike. The **direction of dip** is the direction of maximum slope (or the direction a ball would run over the bed if its surface were perfectly flat). The **angle of dip** is the acute angle this direction makes with a horizontal plane.

Dip-strike symbols are used in most geologic maps to record strike and dip measurements. The generally used symbol has a long line oriented parallel to the compass direction of the strike. A short tick line is placed in the center of the line on the side to which the inclined plane dips, and the angle of dip is recorded next to the strike and dip symbols as shown in figure 5.6. For beds with a ninety-degree vertical dip the short line crosses the strike line, and for beds with no horizontal dip a circle with a cross inside is used, as shown in figure 5.6.

Faults—the cause of earthquakes—involve brittle rock fractures and relative movement of rock units. In nearly all cases, the rocks involved were originally in a horizontal position. The amount of movement may vary from less than a few inches to many thousands of feet vertically and to more than 100 miles horizontally. Different types of faults are produced by different compressional and tensional stresses, and they also depend upon the rock type and geological setting.

## TYPES OF FAULTS

Depending on the direction of relative displacement, faults can be divided into several different types. Because faults are planar features, the concept of strike and dip also applies, and thus the strike and dip of a fault plane can be measured. Types of faults are discussed below.

Strike and dip symbol for vertical beds

Strike and dip symbol for horizontal beds

**Figure 5.6.   Strike and dip symbols for geological maps.**

- **Dip slip faults**—have an inclined fault plane and along which the relative displacement or offset has occurred along the dip direction.

## Important Point!

In looking at the displacement of any fault, we don't know which side actually moved or if both sides moved; all we can determine is the relative sense of motion.

The block above the fault, for any inclined plane, is known as the *hanging wall block* and the block below the fault is the *footwall block*.

- **Normal faults**—are faults in which relative downward movement has taken place down the upper face or hanging wall of the fault plane (see figure 5.7).
- **Horsts and grabens**—faults adjacent to normal faults that dip in opposite directions, as faults often occur in a series due to the tensional stress. In this situation the down-dropped fault blocks form *grabens* and the uplifted fault blocks form *horsts* (see figure 5.8). Rift valleys are graben structures hundreds of miles in length. The most prominent is that along the Red Sea, but the basin and range province of Nevada, Utah, and Idaho is also an example. In the basin and range, the basins are elongated grabens that now form valleys, and the ranges are uplifted horst blocks. They also occur below the oceans along the crests of the mid-oceanic ridges.
- **Half-grabens**—these are bounded by only one fault instead of the two that form a normal graben (see figure 5.9).
- **Reverse faults**—have relative upward movement of the hanging wall of the fault plane (see figure 5.10). They occur in areas of horizontal compressional stresses and folding such as mountain belts.

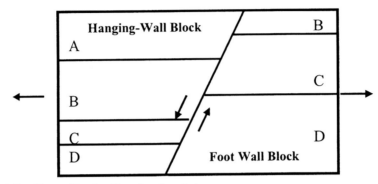

Figure 5.7.   Normal fault with extensional stress.

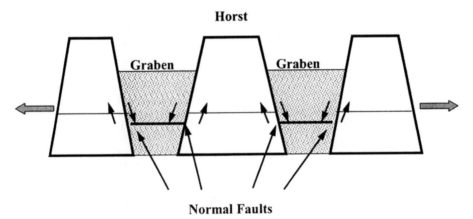

Figure 5.8.   Horst and grabens.

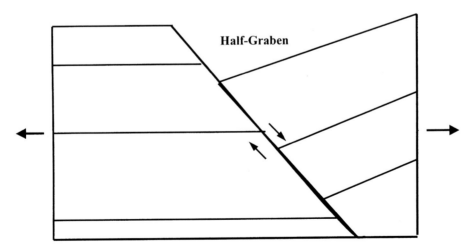

Figure 5.9.   Half-graben.

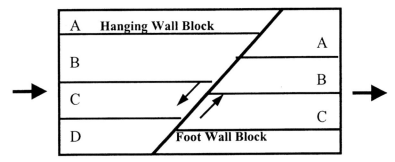

**Figure 5.10.   Reverse fault compressional stress.**

- **Thrust fault**—Chief Mountain in Montana is an example of a special case of a reverse fault where the dip of the fault is less than fifteen degrees (low dip). Thrust faults can have considerable displacement, measuring hundreds of miles, and can result in older strata overlying younger strata (see figure 5.11).
- **Strike slip faults (tear or transform faults)**—are those where shearing stress has produced horizontal movement. The San Andreas Fault (of 1906 San Francisco fame) is an example of a strike-slip fault (see figure 5.12), or more precisely an example of a transform fault (two horizontal plates that slide past one another in a horizontal manner).

## MOVEMENTS ON FAULTS

Except in desert areas, quarries, and cliff faces, faults are rarely seen at the surface, but their presence is indicated by one or more of the following features:

- **Fault breccias**—occur where the ground-up rocks of the fault zone are scattered into angular, irregular-sized, crumbled-up rock fragments.
- **Slikensides**—are polished striations or flutings (scratch marks) that are left on the fault plane as one block moves relative to the other. They are often found in the fault zone.

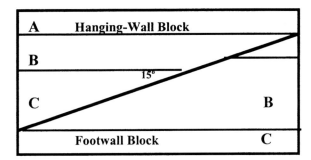

**Figure 5.11.   Thrust fault compressional stress.**

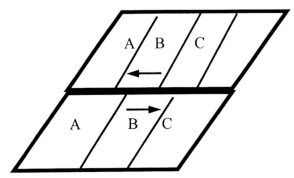

**Figure 5.12.   Strike-slip fault (left-lateral).**

# Folding of Rocks

Folds are wrinkles or flexures in stratified, ductile rocks. Folds result from compressional stresses acting over considerable time. They sometimes occur in isolation, but especially in mountain ranges they are more often packed together. Upfolds (where the originally horizontal strata have been folded upward) are called *anticlines* (see figure 5.13), and downfolds (where the two limbs [sides] of the fold dip inward toward the hinge of the fold) are called *synclines* (see figure 5.14).

### CLASSIFICATION OF FOLDS

Folds can be classified based on their appearance.

- **Symmetrical folds**—the axial plane is vertical with limbs dipping in opposite directions at the same inclinations (angles).

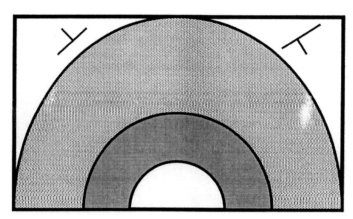

**Figure 5.13.   Anticline.**

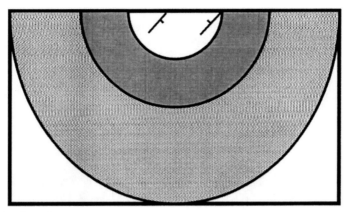

**Figure 5.14.    Syncline.**

- **Asymmetrical folds**—are those having an inclined axial plane and, like symmetrical folds, have limbs dipping in opposite directions, but at different inclinations.
- **Overturned folds**—these occur when the folding is so intense that the strata on one limb of the fold becomes nearly upside down.
- **Isoclinal folds** (*iso* means "same," and *cline* means "angle," so *isoclinal* means the limbs have the same angle)—these occur when compressional stresses that cause the folding are intense, causing the fold to close up and have limbs that are parallel to each other.
- **Recumbent fold**—these occur when an overturned fold has an axial plant that is nearly horizontal.
- **Chevron fold**—these are folds that have no curvature in their hinge and straight-side limbs that form a zigzag pattern.

## Did You Know?

All rocks do not behave the same under stress. When subjected to the same stress, some rocks will fracture or fault, while others will fold. When such contrasting rocks occur in the same area, such as brittle rocks overlaid by ductile rocks, the ductile rocks may bend or fold over the fault while the brittle rocks may fault.

# Deformation of the Crust and Mountain Building

One of the most conspicuous and spectacular results of deformation acting within the crust of the earth is the formation of mountain ranges. Any isolated, upstanding mass can be called a mountain. There is no minimum height or shape involved. There are three types of mountains, and they originate by three processes (two of which are related to deformation):

1. **Fault block mountains**—as the name implies, fault block mountains originate by faulting—normal and reverse faulting—in any type of rock. The Sierra Nevada represents an uplifted block of granite 400 miles long and 100 miles wide.
2. **Fold and thrust mountains**—these mountains all have similar characteristics. They are the result of large compressional forces that cause continental crustal areas to collide, resulting in folding and thrusting of crustal blocks. The Alps, the Himalayas, and the Appalachian Mountains were formed by such processes.
3. **Volcanic mountains**—include some of the most spectacular and beautiful mountains. Mount Rainier and Mount St. Helens in Washington State, Mount Vesuvius in Italy, Mount Hood in Oregon, and the mountains of the Hawaiian Islands and Iceland are all characteristically steep and symmetrical volcanic mountains.

## Chapter Review Questions

5.1   Metamorphism occurs because some minerals are stable only under certain conditions of _____ and _____.

5.2   Limestone is formed by the mineral _____.

5.3   A _____ is a body of rock with specified characteristics.

5.4   Stress that stretches rock: _____.

5.5   Stress that result in slippage and translation: _____.

5.6   Uplifted fault blocks form _____.

5.7   Polished striations or flutings (scratch marks) that are left on the fault plane as one block moves relative to the other: _____.

5.8   _____ can be classified based on their appearance.

5.9   _____ folds have no curvature in their hinge and straight-side limbs that form a zigzag pattern.

5.10  Deformation of the crust equals _____ _____.

## References and Recommended Reading

Dinwiddie, R., et al. 2005. *Universe*. New York: DK Publishing/Penguin Books.
Dinwiddie, R., et al. 2003. *Earth*. New York: DK Publishing/Penguin Books.

## CHAPTER 6

# Weathering and Soil Formation

In regards to surface rock formations, weathering and erosion are both creators and executioners! As soon as rock is lifted above sea level, weather starts to break it up. Water, ice, and chemicals split, dissolve, or rot the rocky surface until it crumbles. Mixed with water and air and plant and animal remains, crumbled rock forms soil.

## Weathering

One can't obtain even the slimmest edge of understanding of geology without understanding the processes that cause the breakdown of rocks, either to form new minerals that are stable on the surface of the earth, or to break the rocks down to smaller particles. Simply, weathering (which projects itself on all surface material above the water table) is the general term used for all the ways in which a rock may be broken down.

### FACTORS THAT INFLUENCE WEATHERING

The factors that influence weathering include:

- **Rock type and structure**—each mineral contained in rocks has a different susceptibility to weathering. A rock with bedding planes, joints, and fractures provides pathways for the entry of water, leading to more rapid weathering. Differential weathering (rocks erode at differing rates) can occur when rock combinations consist of rocks that weather faster than more resistant rocks.
- **Slope**—on steep slopes weathering products may be quickly washed away by rains. Wherever the force of gravity is greater than the force of friction holding particles upon a slope, these tend to slide downhill.
- **Climate**—when higher temperatures and high amounts of water generally cause chemical reactions to run faster. Rates of weathering are higher in warmer than in colder dry climates.
- **Animals**—rodents, earthworms, and ants that burrow into soil bring material to the surface where it can be exposed to the agents of weathering.
- **Time**—depends on slope, animals, and climate.

### CATEGORIES OF WEATHERING PROCESSES

Although weathering processes are separated, it is important to recognize that these processes work in tandem to break down rocks and minerals to smaller fragments. Geologists recognize two categories of weathering processes

1. **Physical (or mechanical) weathering**—disintegration of rocks and minerals by a physical or mechanical process.
2. **Chemical weathering**—the decomposition of rock by chemical changes or solution.

*Physical Weathering*

Physical weathering involves the disintegration of a rock by physical processes. These include freezing and thawing of water in rock crevices, disruption by plant roots or burrowing animals, and the changes in volume that result from chemical weathering with the rock. These and other physical weathering processes are discussed below.

- **Development of joints**—joints are another way that rocks yield to stress. Joints are fractures or cracks in which the rocks on either side of the fracture have not undergone relative movement. Joints form as a result of expansion due to cooling or relief of pressure as overlying rocks are removed by erosion. They form free space in rock by which other agents of chemical or physical weathering can enter (unlike faults that show offset across the fracture). They play an important part in rock weathering as zones of weakness and water movement.
- **Crystal growth**—as water percolates through fractures and pore spaces it may contain ions that precipitate to form crystals. When crystals grow they can cause the necessary stresses needed for mechanical rupturing of rocks and minerals.
- **Heat**—it was once thought that daily heating and cooling of rocks was a major contributor to the weathering process. This view is no longer shared by most practicing geologists. However, it should be pointed out that sudden heating of rocks from forest fires may cause expansion and eventual breakage of rock
- **Biological activities**—plant and animal activities are important contributors to rock weathering. Plants contribute to the weathering process by extending their root systems into fractures and growing, causing expansion of the fracture. Growth of plants and their effects are evident in many places where they are planted near cement work (streets, brickwork, and sidewalks). Animal burrowing in rock cracks can break rock.
- **Frost Wedging**—is often produced by alternate freezing and thawing of water in rock pores and fissures. Expansion of water during freezing causes the rock to fracture. Frost wedging is more prevalent at high altitudes where there may be many freeze-thaw cycles. One classic and striking example of weathering of Earth's surface rocks by frost wedging is illustrated by the formation of hoodoos in Bryce Canyon National Park, Utah (see figure 6.1). "Although Bryce Canyon receives a meager 18 inches of precipitation annually, it's amazing what this little bit of water can do under the right circumstances" (NPS, 2008).

  Approximately 200 freeze-thaw cycles occur annually in Bryce. During these periods, snow and ice melt in the afternoon and water seeps into the joints of the Bryce or Claron Formation. When the sun sets, temperatures plummet and the water refreezes, expanding up to 9 percent as it becomes ice. This frost wedging process exerts tremendous pressure or force on the adjacent rock and shatters and

**Figure 6.1.    Frost-wedged-formed hoodoos. Bryce Canyon National Park, Utah. Photo by Frank R. Spellman.**

pries the weak rock apart. The assault from frost wedging is a powerful force but, at the same time, rain water (the universal solvent), which is naturally acidic, slowly dissolves away the limestone, rounding off the edges of these fractured rocks, and washing away the debris. Small rivulets of water round down Bryce's rime, forming gullies. As gullies are cut deeper, narrow walls of rock known as fins being to emerge. Fins eventually develop holes known as windows. Windows grow larger until their roofs collapse, creating hoodoos (see figure 6.1). As old hoodoos age and collapse, new ones are born (NPS, 2008).

## Did You Know?

Bryce Canyon National Park lies along the high eastern escarpment of the Paunsaugunt Plateau in the Colorado Plateau region of southern Utah. Its extraordinary geological character is expressed by thousands of rock chimneys (hoodoos) that occupy amphitheater-like alcoves in the Pink Cliffs, whose bedrock host is Claron Formation of Eocene age (Davis and Pollock 2003).

# Did You Know?

Hoodoo Pronunciation: "HU-du"
Noun
Etymology: West African; from voodoo
A natural column of rock in western North America often in fantastic form.
—Merriam-Webster Online (www.m-w.com)

## Chemical Weathering

Chemical weathering involves the decomposition of rock by chemical changes or solution. Rocks that are formed under conditions present deep within the earth are exposed to conditions quite different (i.e., surface temperatures and pressures are lower on the surface and copious amounts of free water and oxygen are available) when uplifted onto the surface. The chief processes are oxidation, carbonation and hydration, and solution in water above and below the surface.

## The Persistent Hand of Water

Because of its unprecedented impact on shaping and reshaping Earth, at this point in the text, it is important to point out that given time, nothing, absolutely nothing on Earth is safe from the heavy hand of water. The effects of water sculpting by virtue of movement and accompanying friction will be covered later in the text. For now, in regards to water exposure and chemical weathering, the main agent responsible for chemical weathering reactions is not water movement but instead is weak acids formed in water.

The acids formed in water are solutions that have abundant free hydrogen$^+$ ions. The most common weak acid that occurs in surface waters is carbonic acid. Carbonic acid ($H_2CO_3$) is produced when atmospheric carbon dioxide dissolves in water; it exists only in solution. Hydrogen ions are quite small and can easily enter crystal structures, releasing other ions into the water.

$$\underset{\text{water}}{H_2O} + \underset{\text{carbon dioxide}}{CO_2} \rightarrow \underset{\text{carbonic acid}}{H_2CO_3} \rightarrow \underset{\text{hydrogen ion}}{H^+} + \underset{\text{bicarbonate ion}}{HCO_3^-}$$

## Types of Chemical Weathering Reactions

As mentioned, chemical weathering breaks rocks down by adding or removing chemical elements and changing them into other materials. Again, as stated, chemical weathering consists of chemical reactions, most of which involve water. Types of chemical weathering include:

• **Hydrolysis**—this is a water-rock reaction that occurs when an ion in the mineral is replaced by $H^+$ or $OH^-$.

- **Leaching**—ions are removed by dissolution into water.
- **Oxidation**—oxygen is plentiful near Earth's surface; thus, it may react with minerals to change the oxidation state of an ion.
- **Dehydration**—occurs when water or a hydroxide ion is removed from a mineral.
- **Complete dissolution**

# Chapter Review Questions

6.1  The factors that influence weathering include: _____.
6.2  Disintegration of rocks and minerals by a physical or mechanical process: _____.
6.3  Involves the decomposition of rock by chemical changes or solution: _____.
6.4  Alternate freezing and thawing of water in rock pores and fissures: _____.
6.5  A common weak acid in surface waters: _____.
6.6  _____ cover bare rock first.

# References and Recommended Reading

ASTM. 1969. *Manual on Water*. Philadelphia: American Society for Testing and Materials.

Beazley, J. D. 1992. *The Way Nature Works*. New York: Macmillan Publishing Company.

Brady, N. C., and Weil, R. R. 1996. *The Nature and Properties of Soils*, 11th ed. Upper Saddle River, NJ: Prentice-Hall.

Carson, R. 1962. *Silent Spring*. Boston: Houghton Mifflin Company.

Ciardi, J. 1997. Stoneworks. In *The Collected Poems of John Ciardi*, ed. E. M. Cifelli. Fayettville: University of Arkansas Press.

Davis, G. H., and Pollock, G. L. 2003. Geology of Bryce Canyon National Park, Utah. In *Geology of Utah's Parks and Monuments*, 2nd ed., ed. D. A. Sprinkel et al. Salt Lake City: Utah Geological Association.

Eswaran, H. 1993. Assessment of global resources: Current status and future needs. *Pedologie* 43, pp. 19–39.

Foth, H. D. 1992. *Fundamentals of Soil Science*, 6th ed. New York: Wiley.

Franck, I., and Brownstone, D. 1992. *The Green Encyclopedia*. New York: Prentice-Hall.

Kemmer, F. N. 1977. *Water: The Universal Solvent*. Oak Ridge, IL: NALCO Chemical Company.

Konigsburg, E. M. 1996. *The View from Saturday*. New York: Scholastic.

NPS. 2008. *The Hoodoo*. Washington, DC: National Park Service.

Spellman, F. R. 1998. *The Science of Environmental Pollution*. Boca Raton, FL: CRC Press.

Mowat, F. 1957. *The Dog Who Wouldn't Be*. New York: Willow Books.

USDA. 1975. *Soil Survey Staff, Soil Classification, A Comprehensive System*. Washington, D.C.: USDA Natural Resources Conservation Service.

Soil Survey Staff, Keys to Soil Taxonomy. 1994. Washington, D.C.: USDA Natural Resources Conservation Service.

Tomera, A. N. 1989. *Understanding Basic Ecological Concepts*. Portland, ME: J. Weston Walch, Publisher.

USDA. 1975. *Soil Taxonomy: A Basic System of Soil Classification for Making and Interpreting Soil Surveys*. Washington, D.C.: USDA Natural Resources Conservation Service.

CHAPTER 7

# Stream and Groundwater Systems

## Stream Genesis and Structure

Early in the spring on a snow- and ice-covered high alpine meadow, the process known as the water cycle resumes after the long dark winter months. The cycle's main component, water, has been held in reserve—literally frozen. But with longer, warmer spring days, the sun is higher, more direct, and of longer duration, and the frozen masses of water respond to the increased warmth. The melt begins with a single drop, then two, then more and more. As the snow and ice melts, the drops join a chorus that continues unending; they fall from their ice-bound lip to the bare rock and soil terrain below.

Unlike glacial till—the unconsolidated, heterogeneous mixture of clay, sand, gravel, and boulders—the terrain that the snowmelt strikes was not dug out, ground out, and exposed by the force of a huge, slow and inexorably moving glacier. Instead, this soil and rock is exposed because of a combination of wind and the tiny, enduring force exerted by drops of water as season after season they collide with the thin soil cover, exposing the intimate bones of the earth.

Gradually, the single drops increase to separate rivulets that trickle and then rush, joining to form a splashing, rebounding, helter-skelter cascade down the face of the granite mountain. At an indented ledge halfway down the mountain slope, a pool forms, whose beauty, clarity, and sweet iciness provide the visitor with an incomprehensible, incomparable gift—a blessing from the earth.

The mountain pool fills slowly, tranquil under the blue sky, reflecting the pines, snow, and sky around and above it, an open invitation to lie down and drink and to peer into the glass-clear, deep waters, so clear that it seems possible to reach down over fifty feet and touch the very bowels of the mountain. The pool has no transition from shallow margin to depth; it is simply deep and pure. As the pool fills with more meltwater, we wish to freeze time, to hold this place and this pool in its perfect state forever; it is such a rarity to us in our modern world. But this cannot be—Mother Nature calls, prodding, urging—and for a brief instant, the water laps against the outermost edge of the ridge, then a trickle flows over the rim. The giant hand of gravity reaches out and tips the overflowing melt onward, and it continues its journey down, following the path of least resistance to its next destination, several thousand feet below.

When the overflow, still high in altitude, meets the angled, broken rocks below, it bounces, bursts, and mists its way against steep, V-shaped walls that form a small valley, carved out over time by water and the forces of the earth.

Within the valley confines, the meltwater has grown from drops to rivulets to a

103

small mass of flowing water. It flows through what is at first a narrow opening, gaining strength, speed, and power as the V-shaped valley widens to form a U shape. The journey continues as the water mass picks up speed and tumbles over massive boulders, and then slows again.

At a larger but shallower pool, other waters from higher elevations have joined the main body—from the hillsides, crevices, springs, rills, mountain creeks. At the influent pool sides, all appears peaceful, quiet, and restful, but not far away, at the effluent end of the pool, gravity takes control again. The overflow is flung over the jagged lip, and cascades downward several hundred feet to a violent, mist-filled meeting.

The water separates and joins again and again, forming a deep, furious, wild stream that calms gradually as it continues to flow over lands that are less steep. The waters widen into pools overhung by vegetation, surrounded by tall trees. The pure, crystalline waters have become progressively discolored on their downward journey, stained brown-black with humic acid and literally filled with suspended sediments; the once-pure stream is now muddy.

The mass divides and flows in different directions, over different landscapes. Small streams divert and flow into open country. Different soils work to retain or speed the waters, and in some places the waters spread out into shallow swamps, bogs, marshes, fens, or mires. Other streams pause long enough to fill deep depressions in the land and form lakes. For a time, the water remains and pauses in its journey to the sea. But this is only a short-term pause, because lakes are only a short-term resting place in the water cycle. The water will eventually move on, by evaporation or seepage into groundwater. Other portions of the water mass stay with the main flow, and the speed of flow changes to form a river, which braids its way through the landscape, heading for the sea. As the river changes speed and slows, the river bottom changes from rock and stone to silt and clay. Plants begin to grow, stems thicken, and leaves broaden. The river is now full of life and the nutrients needed to sustain life. But the river courses onward, its destiny met when the flowing rich mass slows its last and finally spills into the sea (Spellman and Whiting, 1998).

# Streams

Streams are bodies of running water that carry rock particles (sediment loads) and dissolved ions and flow down slope along a clearly defined path, called a *channel*. Thus, streams may vary in width from a few inches to several miles. Streams are important for several reasons:

- Streams are an important part of the water cycle; they carry most of the water that goes from the land to the sea.
- Streams are one of the main transporters of sediment load from higher to lower elevations.
- Streams carry dissolved ions, the products of chemical weathering, into the oceans and thus make the sea salty.

- Streams (along with weathering and mass wasting) are a major part of the erosional process.
- Most population centers are located next to streams because they provide a major source of water and transportation.

# Key Terms

**Evapotranspiration (plant water loss)**—describes the process whereby plants lose water to the atmosphere during the exchange of gases necessary for photosynthesis. Water loss by evapotranspiration constitutes a major flux back to the atmosphere.

**Infiltration Capacity**—the maximum rate soil can absorb rainfall.

**Perennial stream**—a type of stream in which flow continues during periods of no rainfall.

**Gaining stream**—typical of humid regions, where groundwater recharges the stream.

**Losing stream**—typical of arid regions, where streams can recharge groundwater.

**Laminar flow**—in a stream where parallel layers of water shear over one another vertically.

**Turbulent flow**—in a stream where complex mixing is the result.

**Meandering**—stream condition whereby flow follows a winding and turning course.

**Thalweg**—line of maximum water of channel depth in a stream.

**Riffles**—refer to shallow, high-velocity flow over mixed gravel-cobble (bar-like) substrate.

**Sinuosity**—the bending or curving shape of a stream course.

# Characteristics of Stream Channels

A standard rule of thumb states: Flowing waters (rivers and streams) determine their own channels, and these channels exhibit relationships attesting to the operation of physical laws—laws that are not, as of yet, fully understood. The development of stream channels and entire drainage networks, and the existence of various regular patterns in the shape of channels, indicate that streams are in a state of dynamic equilibrium between erosion (sediment loading) and deposition (sediment deposit), and governed by common hydraulic processes. However, because channel geometry is four dimensional, with a long profile, cross-section, depth, and slope profile, and because these mutually adjust over a time scale as short as years and as long as centuries or more, cause-and-effect relationships are difficult to establish. Other variables that are presumed to interact as the stream achieves its graded state include width and depth, velocity, size of sediment load, bed roughness, and the degree of braiding (sinuosity).

## STREAM PROFILES

Mainly because of gravity, most streams exhibit a downstream decrease in gradient along their length. Beginning at the headwaters, the steep gradient becomes less so as one proceeds downstream, resulting in a concave longitudinal profile. Though diverse geography provides for almost unlimited variation, a lengthy stream that originates in a mountainous area (such as the one described in the chapter opening) typically comes into existence as a series of springs and rivulets; these coalesce into a fast-flowing, turbulent mountain stream, and the addition of tributaries results in a large and smoothly flowing river that winds through the lowlands to the sea.

When studying a stream system of any length, it becomes readily apparent (almost from the start of such studies) that what we are studying is a body of flowing water that varies considerably from place to place along its length. For example, a common variable—the results of which can be readily seen—is increase or decrease of discharge, causing corresponding changes in the stream's width, depth, and velocity. In addition to physical changes that occur from location to location along a stream's course, there are myriad biological variables that correlate with stream size and distance downstream. The most apparent and striking changes are in steepness of slope and in the transition from a shallow stream with large boulders and a stony substrate to a deep stream with a sandy substrate.

The particle size of bed material at various locations is also variable along the stream's course. The particle size usually shifts from an abundance of coarser material upstream to mainly finer material in downstream areas.

## SINUOSITY

Unless forced by man in the form of heavily regulated and channelized streams, straight channels are uncommon. Stream flow creates distinctive landforms composed of straight (usually in appearance only), meandering, and braided channels, channel networks, and flood plains. Simply put: flowing water will follow a sinuous course. The most commonly used measure is the **sinuosity index** (SI). Sinuosity equals 1 in straight channels and more than 1 in sinuous channels.

$$SI = \frac{\text{Channel distance}}{\text{Down valley distance}}$$

Meandering is the natural tendency for alluvial channels and is usually defined as an arbitrarily extreme level of sinuosity, typically an SI greater than 1.5. Many variables affect the degree of sinuosity, however, so SI values range from near unity in simple, well-defined channels to 4 in highly meandering channels (Gordon et al., 1992).

It is interesting to note that even in many natural channel sections of a stream course that appear straight, meandering occurs in the line of maximum water or channel depth (known as the thalweg). Keep in mind that a stream has to meander; that is how they renew themselves. By meandering, they wash plants and soil from the land into their waters, and these serve as nutrients for the plants in the rivers. If rivers aren't allowed to meander, if they are channelized, the amount of life they can support will gradually decrease. That means less fish, ultimately—and less bald eagles, herons, and other fishing birds (Spellman, 1996).

Meander flow follows a predictable pattern and causes regular regions of erosion and deposition. The streamlines of maximum velocity and the deepest part of the channel lie close to the outer side of each bend and cross over near the point of inflection between the banks. A huge elevation of water at the outside of a bend causes a helical flow of water toward the opposite bank. In addition, a separation of surface flow causes a back eddy. The result is zones of erosion and deposition, and this explains why point bars develop in a downstream direction in depositional zones (Morisawa, 1968).

## Did You Know?

Meandering channels can be highly convoluted or merely sinuous, but all maintain a single thread, with curves having definite geometric shape. Straight channels are sinuous but apparently random in occurrence of bends. Braided channels are those with multiple streams separated by bars.

## BARS, RIFFLES, AND POOLS

Implicit in the morphology and formation of meanders are bars, riffles, and pools. **Bars** develop by deposition in slower, less competent flow on either side of the sinuous main stream. Onward-moving water, depleted of bed load, regains competence and shears a pool in the meander—reloading the stream for the next bar. Alternating bars migrate to form riffles.

As stream flow continues along its course, a pool-riffle sequence is formed. Basically the **riffle** is a mound or hillock and the **pool** is a depression.

## FLOODPLAIN

Stream channels influence the shape of the valley floor through which they course. This self-formed, self-adjusted flat area near to the stream is the floodplain, which loosely describes the valley floor, prone to periodic inundation during over-bank discharges. What is not commonly known is that valley flooding is a regular and natural behavior of the stream. Many people learn about this natural phenomenon the hard way—that is, whenever their farms, towns, streets, and homes become inundated by a river or stream that is doing nothing more than following its "natural" periodic cycle—conforming to the Master Plan designed by the Master Planner: Mother Nature.

## Did You Know?

Floodplain rivers are found where regular floods form lateral plains outside the normal channel that seasonally become inundated, either as a consequence of greatly increased rainfall or snowmelt.

# Water Flow in a Stream

Most elementary students learn early in their education process that water on Earth flows downhill (gravity)—from land to the sea. However, they may or may not be told that water flows downhill toward the sea by various routes.

For the moment, the "route" (channel, conduit, or pathway) we are concerned with is the surface water route taken by surface runoff. Surface runoff is dependent on various factors. For example, climate, vegetation, topography, geology, soil characteristics, and land use determine how much surface runoff occurs compared with other pathways.

The primary source (input) of water to total surface runoff, of course, is precipitation. This is the case even though a substantial portion of all precipitation input returns directly to the atmosphere by evapotranspiration. Evapotranspiration is a combination process, as the name suggests, whereby water in plant tissue and in the soil evaporates and transpires to water vapor in the atmosphere.

Probably the easiest way to understand precipitation's input to surface water runoff is to take a closer look at this precipitation input.

Again, a substantial portion of precipitation input returns directly to the atmosphere by evapotranspiration. It is also important to point out that when precipitation occurs, some rainwater is intercepted or blocked or caught by vegetation, where it evaporates, never reaching the ground or being absorbed by plants. A large portion of the rainwater that reaches the surface on ground and in lakes and streams also evaporates directly back to the atmosphere. Although plants display a special adaptation to minimize transpiration, plants still lose water to the atmosphere during the exchange of gases necessary for photosynthesis. Notwithstanding the large percentage of precipitation that evaporates, rain- or meltwater that reaches the ground surface follows several pathways in reaching a stream channel or groundwater.

Soil can absorb rainfall to its infiltration capacity (i.e., to its maximum intake rate). During a rain event, this capacity decreases. Any rainfall in excess of infiltration capacity accumulates on the surface. When this surface water exceeds the depression storage capacity of the surface, it moves as an irregular sheet of overland flow. In arid areas, overland flow is likely because of the low permeability of the soil. Overland flow is also likely when the surface is frozen and/or when human activities have rendered the land surface less permeable. In humid areas, where infiltration capacities are high, overland flow is rare.

In rain events where the infiltration capacity of the soil is not exceeded, rain penetrates the soil and eventually reaches the groundwater—from which it discharges to the stream slowly and over a long period of time. This phenomenon helps to explain why stream flow through a dry-weather region remains constant; the flow is continuously augmented by groundwater. This type of stream is known as a perennial stream, as opposed to an intermittent one, because the flow continues during periods of no rainfall.

Streams that course their way in channels through humid regions are fed water via the water table, which slopes toward the stream channel. Discharge from the water table into the stream accounts for flow during periods without precipitation and also explains why this flow increases, even without tributary input, as one proceeds downstream. Such streams are called **gaining**, or **effluent**, as opposed to **losing**, or **influent** streams that lose water into the ground. It is interesting to note that the same stream can shift between gaining and losing conditions along its course because of changes in underlying strata and local climate.

# Stream Water Discharge

The current velocity (speed) of water (driven by gravitational energy) in a channel varies considerably within a stream's cross-section, owing to friction with the bottom and sides, with sediment, with obstructions (rocks and logs, etc.), and with the atmosphere, and to sinuosity (bending or curving). Highest velocities, obviously, are found where friction is least, generally at or near the surface and near the center of the channel. In deeper streams, current velocity is greatest just below the surface, due to the friction with the atmosphere; in shallower streams current velocity is greatest at the surface, due to friction with the bed. Velocity decreases as a function of depth, approaching zero at the substrate surface. A general and convenient rule of thumb is that the deepest part of the channel occurs where the stream velocity is the highest. Additionally, both width and depth of a stream increase downstream because discharge (the amount of water passing any point in a given time) increases downstream. As discharge increases, the cross-sectional shape will change, with the stream becoming deeper and wider. Velocity is important to discharge because discharge ($m^3$/sec) = cross-sectional Area (width $\times$ average depth) ($m^2$) $\times$ Average Velocity (m/sec).

$$Q = A \times V$$

A stream is constantly seeking balance. This can be seen whenever the amount of water in a stream increases, the stream must adjust its velocity and cross-sectional area to reach balance. Discharge increases as more water is added through precipitation, tributary streams, or from groundwater seeping into the stream. As discharge increases, generally width, depth, and velocity of the stream also increase.

# Transport of Material (Load)

Water flowing in a channel may exhibit **laminar flow** (parallel layers of water shear over one another vertically), or **turbulent flow** (complex mixing) (see figure 7.1). In

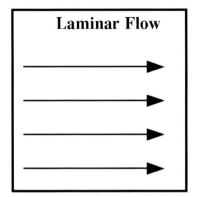

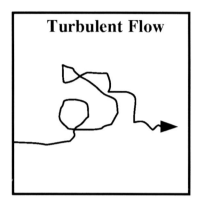

**Figure 7.1. Laminar and turbulent flow.**

streams, laminar flow is uncommon, except at boundaries where flow is very low and in groundwater. Thus the flow in streams generally is turbulent. Turbulence exerts a shearing force that causes particles to move along the streambed by pushing, rolling, and skipping, referred to as bed load. This same shear causes turbulent eddies that entrain particles in suspension (called the suspended load—particles size under 0.06 millimeters). **Entrainment** is the incorporation of particles when stream velocity exceeds the entraining velocity for a particular particle size.

## Did You Know?

Entrainment is a natural extension of erosion and is vital to the movement of stationary particles in changing flow conditions. Remember, all sediments ultimately derive from erosion of basin slopes, but the immediate supply usually derives from the stream channel and banks, while the bedload comes from the streambed itself and is replaced by erosion of bank regions.

The entrained particles in suspension (suspended load) also include fine sediment, primarily clays, silts, and fine sands that require only low velocities and minor turbulence to remain in suspension. These are referred to as wash load (under 0.002 millimeters), because this load is "washed" into the stream from banks and upland areas (Gordon et al., 1992; Spellman, 1996).

Thus the suspended load includes the wash load and coarser materials (at lower flows). Together, the suspended load and bed load constitute the solid load. It is important to note that in bedrock streams the bed load will be a lower fraction than in alluvial streams where channels are composed of easily transported material.

A substantial amount of material is also transported as the dissolved load. Solutes (ions) are generally derived from chemical weathering of bedrock and soils, and their contribution is greatest in subsurface flows and in regions of limestone geology.

The relative amount of material transported as solute rather than solid load depends on basin characteristics, lithology (i.e., the physical character of rock), and hydrologic pathways. In areas of very high runoff, the contribution of solutes approaches or exceeds sediment load, whereas in dry regions, sediments make up as much as 90 percent of the total load.

Stream competence refers to the largest particle that can be moved as bed load and the lowest velocity at which a particle resting on the streambed will move. Deposition occurs when stream competence falls below a given velocity. Simply stated: *the size of the particle that can be eroded and transported is a function of current velocity.*

Sand particles are the most easily eroded. The greater the mass of larger particles (e.g., coarse gravel), the higher the initial current velocities must be for movement. However, smaller particles (silts and clays) require even greater initial velocities because of their cohesiveness and because they present smaller, streamlined surfaces to the flow. Once in transport, particles will continue in motion at somewhat slower velocities than initially required to initiate movement, and will settle at still lower velocities.

Particle movement is determined by size, flow conditions, and mode of entrainment. Particles over 0.02 millimeters (medium-coarse sand size) tend to move by rolling or sliding along the channel bed as traction load. When sand particles fall out of the flow, they move by saltation or repeated bouncing. Particles under 0.06 millimeters (silt) move as suspended load, and particles under 0.002 millimeters (clay), indefinitely, as wash load. A considerable amount of particle sorting takes place because of the different styles of particle flow in different sections of the stream (Richards, 1982; Likens, 1984).

Unless the supply of sediments becomes depleted, the concentration and amount of transported solids increases. However, discharge is usually too low, throughout most of the year, to scrape or scour, shape channels, or move significant quantities of sediment in all but sand-bed streams, which can experience change more rapidly. During extreme events, the greatest scour occurs and the amount of material removed increases dramatically.

Sediment inflow into streams can both be increased and decreased as a result of human activities. For example, poor agricultural practices and deforestation greatly increase erosion. Man-made structures such as dams and channel diversions can, on the other hand, greatly reduce sediment inflow.

# Groundwater

Unbeknownst to most of us, Earth possesses an unseen ocean of water. This ocean, unlike the surface oceans that cover most of the globe, is freshwater: the groundwater that lies contained in aquifers beneath Earth's crust. This gigantic freshwater water source (about 1 percent of the water on Earth; about thirty-five times the amount of water in lakes and streams) forms a reservoir that feeds all the natural fountains and springs of Earth. But how does water travel into the aquifers that lie under Earth's surface?

Groundwater sources are replenished from a percentage of the average approximately three feet of water that falls to earth each year on every square foot of land.

Water falling to earth as precipitation follows three courses. Some runs off directly to rivers and streams (roughly six inches of that three feet), eventually working back to the sea. Evaporation and transpiration through vegetation takes up about two feet. The remaining six inches seeps into the ground, entering and filling every interstice, each hollow and cavity. Gravity pulls water toward the center of the Earth. That means that water on the surface will try to seep into the ground below it. Although groundwater comprises only one-sixth of the total, (1,680,000 miles of water), if we could spread out this water over the land, it would blanket it to a depth of 1,000 feet.

As mentioned, part of the precipitation that falls on land infiltrates the land surface, percolates downward through the soil under the force of gravity, and becomes groundwater. Groundwater, like surface water, is extremely important to the hydrologic cycle and to our water supplies. Almost half of the people in the United States drink public water from groundwater supplies. Overall, more water exists as ground water than surface water in the United States, including the water in the Great Lakes. But sometimes, pumping it to the surface is not economical, and in recent years, pollution of groundwater supplies from improper disposal has become a significant problem.

We find groundwater in saturated layers called **aquifers** under the earth's surface. Three types of aquifers exist: unconfined, confined, and springs.

Aquifers are made up of a combination of solid material such as rock and gravel and open spaces called pores. Regardless of the type of aquifer, the groundwater in the aquifer is in a constant state of motion. This motion is caused by gravity or by pumping.

The actual amount of water in an aquifer depends upon the amount of space available between the various grains of material that make up the aquifer. The amount of space available is called porosity. The ease of movement through an aquifer is dependent upon how well the pores are connected. For example, clay can hold a lot of water and has high porosity, but the pores are not connected, so water moves through the clay with difficulty. The ability of an aquifer to allow water to infiltrate is called permeability.

The aquifer that lies just under the earth's surface is called the zone of saturation, an unconfined aquifer (see figure 7.2). The top of the zone of saturation is the water table. An **unconfined aquifer** is only contained on the bottom and is dependent on local precipitation for recharge. This type of aquifer is often called a water table aquifer.

Unconfined aquifers are a primary source of shallow well water (see figure 7.2). These wells are shallow (and not desirable as a public drinking water source). They are subject to local contamination from hazardous and toxic materials—fuel and oil, and septic tanks and agricultural runoff providing increased levels of nitrates and microorganisms. These wells may be classified as groundwater under the direct influence of surface water (GUDISW), and therefore require treatment for control of microorganisms.

A **confined aquifer** is sandwiched between two impermeable layers that block the flow of water. The water in a confined aquifer is under hydrostatic pressure. It does not have a free water table (see figure 7.3).

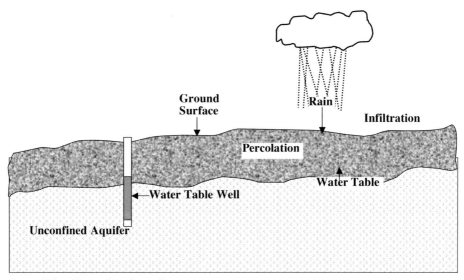

**Figure 7.2.   Unconfined aquifer.**

*Source*: Spellman, F. R. 1996. *Stream Ecology and Self-Purification*. Lancaster, PA: Technomic Publishing Company.

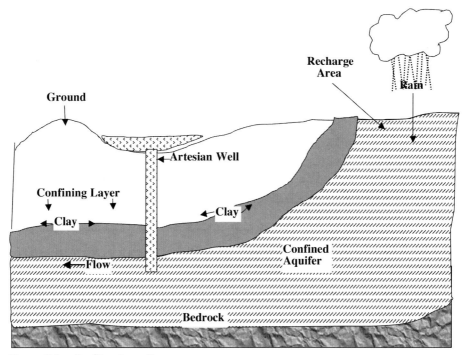

**Figure 7.3.   Confined aquifer.**

*Source*: Spellman, F. R. 1996. *Stream Ecology and Self-Purification*. Lancaster, PA: Technomic Publishing Company.

Confined aquifers are called artesian aquifers. Wells drilled into artesian aquifers are called artesian wells and commonly yield large quantities of high-quality water. An artesian well is any well where the water in the well casing would rise above the saturated strata. Wells in confined aquifers are normally referred to as deep wells and are not generally affected by local hydrological events.

A confined aquifer is recharged by rain or snow in the mountains where the aquifer lies close to the surface of the earth. Because the recharge area is some distance from areas of possible contamination, the possibility of contamination is usually very low. However, once contaminated, confined aquifers may take centuries to recover.

Groundwater naturally exits the earth's crust in areas called springs. The water in a spring can originate from a water table aquifer or from a confined aquifer. Only water from a confined spring is considered desirable for a public water system.

Almost all groundwater is in constant motion through the pores and crevices of the aquifer in which it occurs. The water table is rarely level; it generally follows the shape of the ground surface. Groundwater flows in the downhill direction of the sloping water table. The water table sometimes intersects low points of the ground, where it seeps out into springs, lakes, or streams.

Usual groundwater sources include wells and springs that are not influenced by surface water or local hydrologic events.

As a potable water source, groundwater has several advantages over surface water. Unlike surface water, groundwater is not easily contaminated. Groundwater sources are usually lower in bacteriological contamination than surface waters. Groundwater quality and quantity usually remains stable throughout the year. In the United States, groundwater is available in most locations.

As a potable water source, groundwater does present some disadvantages compared to surface water sources. Operating costs are usually higher because groundwater supplies must be pumped to the surface. Any contamination is often hidden from view. Removing any contaminants is very difficult. Groundwater often possesses high mineral levels, and thus an increased level of hardness, because it is in contact longer with minerals. Near coastal areas, groundwater sources may be subject to saltwater intrusion.

## GEOLOGIC ACTIVITY OF GROUNDWATER

Groundwater contributes to geologic activity in a number of ways, including:

- **Dissolution**—because water is the main agent of chemical weathering (nothing is safe from water), groundwater too is an active weathering agent that in part or in total (limestone) leaches ions from rock.
- **Chemical Cementation and Replacement**—water not only carries ions away from rock but also brings chemical agents into masses of rock or rock structures. When some of the water-transported chemical enters rocks or rocky masses, many act as cement and bind sedimentary rocks together. In a similar water-aided transport

manner, water transports replacement molecules that work to fossilize organic sub-stances or petrify wood.

- **Caves and Caverns**—when large areas of limestone underground are dissolved by the action of groundwater these cavities can become caves or caverns (caves with many interconnected chambers) once the water table is lowered. Once formed, a cave is open to the atmosphere and water percolating in can precipitate new material such as the common cave decorations stalactites (growing and hanging from the ceiling) and stalagmites (growing from the floor upward) (see figure 7.4).
- **Sinkholes**—these are common in areas underlain by limestone, carbonate rock, salt beds, or rocks that can naturally be dissolved by groundwater circulating through them (USGS, 2006).
- **Karst Topography**—in limestone terrains where dissolution is the main type of weathering, groundwater may work to form caves and sinkholes, and their collapse and coalescence may result in this highly irregular topography.

# Chapter Review Questions

7.1    Clearly defined stream path: _____.

7.2    _____ is the process whereby plants lose water to the atmosphere during the exchange of gases necessary for photosynthesis.

**Figure 7.4.    Limestone/calcite formations. Lewis and Clark Caverns, Montana. Photo by Frank R. Spellman.**

7.3  _____ is the maximum rate soil can absorb rainfall.

7.4  _____ is the maximum water of channel depth in a stream.

7.5  _____ describes stream braiding.

7.6  _____ flow follows a predictable pattern and causes regular regions of erosion and deposition.

7.7  A stream is constantly seeking _____.

7.8  Complex mixing in a stream is caused by _____ flow.

7.9  Suspended stream load includes _____ and _____.

7.10  Groundwater is found in saturated layers called _____.

# References and Recommended Reading

Giller, P. S., and Jalmqvist, B. 1998. *The Biology of Streams and Rivers*. Oxford: Oxford University Press.

Gordon, N. D., McMahon, T. A., and Finlayson, B. L. 1992. *Stream Hydrology: An Introduction for Ecologists*. Chichester: Wiley.

Leopold, L. B. 1994. *A View of the River*. Cambridge, MA: Harvard University Press.

Likens, W. M. 1984. Beyond the shoreline: A watershed ecosystem approach. *Vert. Int. Ver. Theor. Aug. Liminol.* 22, pp. 1–22.

Morisawa, M. 1968. *Streams: Their Dynamics and Morphology*. New York: McGraw-Hill.

Richards, K. 1982. *Rivers: Form and Processes in Alluvial Channels*. London: Methuen.

Spellman, F. R. 1996. *Stream Ecology and Self-Purification*. Lancaster, PA: Technomic Publishing Company.

Spellman, F. R., and Whiting, N. 1998. *Environmental Science and Technology*. Rockville, MD: Government Institutes.

USGS. 2006. *Sinkholes*. Accessed 7/06/08 at http://ga.wwater.usgs.gov/edu/earthqwsinkholes.htm.

# CHAPTER 8

# Glaciation

~

## Yurk and Smilodon

He sat on the ground leaning against a deadfall, his leather-wrapped legs pulled tight under him, and watched the swamp. He felt disoriented, detached from the world around him. Even the air around him felt strange; this place was unusually warm. A possum waddled from a copse of vine-maple below him, and Yurk watched the possum move off to the left—the possum, in a hurry, constantly jerking his head to the right, over his shoulder. The possum darted toward the marshy bank, and stopped to sniff the ground. Some noise or an odor carried on the wind seemed suddenly to startle the possum into attention, and he looked back toward Yurk, then moved off into the tall marsh grass, where he disappeared from view.

The rain was coming down in a fine drizzle. The wind sighed through the fir boughs and the afternoon was redolent with tree-perfumed air. Even with the light rain and wind, though, Yurk was warm—warmer than he ever remembered being before. He had never been so warm, his whole body at the same time. By a fire, only what faces the fire is warm.

Yurk's weathered face wore a mesmerized look as he chewed on a piece of bark, resting against the decaying trunk of the fallen tree, almost as if he were un-aware of his surroundings. His eyes glazed over, as if he were there in shell form only—an empty one at that. Maybe his blank state and hypnotized appearance was the result of the view in front of those blank eyes: great truncated tree trunks, black-ened by fire, stood above the surface of the swamp water; stark remnants of a very ancient past. A misty pall hung over the swamp as the blackness lowered over the forest and the swamp took on an eerie, forbidding, spectral quality with the coming of night.

The cry of an owl drifted through the dark forest as Yurk stood (an effort that required much exertion from his tired, ancient body). Carefully he stretched and yawned—careful not because of his frailty, or from a sense of impending danger, but because of instinct—not fear exactly, just instinct. A lifetime, generations of lifetimes, of vigilance for survival (both conscious and unconscious) had taught Yurk to be wary at all times. He was leg-weary and footsore, but that really didn't concern him. He knew his ending time was near—that was why he had traveled better than 200 miles to this place. This place he had come to had been familiar to him years before, but in a very different form. He wanted to see the wonderment of the swampy terrain that lay before him now.

Yurk was viewing something he had heard about from other clan members but had never witnessed before. A swamp.

Yes, a swamp . . . with blackened, truncated tree remnants. In all his life (he was unusually old for his time and circumstances, well over sixty), Yurk had never seen such a sight. Up until now, the landscape he had been familiar with had been covered in snow and ice. He had visited this place many times in the past—what had been a bare plain of ice and snow. He had not been on this journey in many years, but the last time he had come, he had simply trudged through the open area over layers of thick ice and snow. He (and no one else) had any idea that the swamp lay beneath it. In his absence from this place, he had heard the tales from the younger clan hunters, and had decided to take his last journey—to see such a place, such a site— before he died.

It was so warm.

As he stood, wiping his wet brow and looking out upon the swamp, thirty feet to Yurk's left, working toward the top of the craggy ledge on the sheer cliff edge, climbed the cat.

Like Yurk, the cat had come to this place many times in the past; although she could not cognitively determine the exact difference between the past and the present, she, too, knew this place had changed.

It was so warm.

In the past fifteen or so years, the cat, along with her running mates (these cats almost always ventured into the wilderness accompanied—to hunt and to kill required help—sometimes lots of help) had, like Yurk and his clan members, crossed the swamp using the ice-bridge that covered it. But now things were different; the cat knew this. She also knew that something else was different; it was so warm.

The cat (known today as Smilodon or saber-toothed tiger) continued slowly, inexorably up the steep slope of the stony ridge. Climbing to this ledge in the past, she had to allow for the slipperiness of the ice sheet that covered the rock; now she should have had very little difficulty climbing to this high point overlooking the swamp. But now things were different—much different. She was on her last legs, in all ways. But her difficulty was even more than that; even though the going was easier without the ice and snow, she still struggled up to the terminal point. It was so hot. She labored even to breathe.

Yurk and the cat were aware of each other. Each knew the other was there— have no doubt about that. Yurk was probably more fearful of the cat than she was of him . . . but how could anyone tell? The cat preferred feasting on mammoths and mastodons, and Yurk liked that kind of meat himself, so they had been bitter enemies throughout their lives. When confronted with her "only" threat, her only true enemy, the cat knew she was wise to be ready. Life of any sort was difficult enough; not being alert and wary at all times was certainly an invitation to disaster—for both of them.

It was so warm.

But now things were different. Neither the cat nor Yurk was attentive to the other; they were not as alert, as wary of each other as they had been in the past. Each knew, in their own way, that the days of hunting and protecting themselves were behind them—food certainly wasn't a consideration with either. No, food was not a problem; they were not hungry. Afraid? No, not really.

The cat continued her climb and finally reached the summit. She stood looking out upon the swamp (with one eye semifocused on Yurk). Yurk stood below, looking out on the swamp (aware of her presence as well).

They both knew, in their own way that things were different. They could feel the difference; it was so warm.

Warm . . . yes, it was warm. Except for the past few years, during their entire lives, they had never known such warmth, had never seen the snow and ice melt, had never witnessed the swampy landscape now before them. Their world was different . . . fearfully and wonderfully changed.

The warming trend had actually begun about two or three years earlier, though Yurk and the cat had barely been aware of it because the increase in temperatures had been subtle (about a half a degree Fahrenheit every three months or so). But now, now the difference was obvious. The temperature was a least ten degrees warmer than they had ever experienced—thus the melt, the freshly uncovered swamp, the rock-strewn ledge . . . and the warmth, of course.

The cat and Yurk stood for a time, gazing out at the swamp. What this change would mean to their clan and mates—those to follow—they were not capable of determining. What this change would bring to their world, they were not capable of guessing. So they stood, until Yurk sat back down on the ground, his back against the deadfall, and the cat lay down on the heated rocks of the ledge; they were both exhausted, tired, worn out—old, so old . . . and warm, too warm.

About an hour later, as darkness fell total upon the blackened, spectral landscape before them, they both went to sleep . . . the sleep of the dead . . . and their own warmth turned cold.

The ambient temperature continued to rise, even now that it was dark, night. A night that when ended would bring the dawn of a new day . . . and the dawn of a new era.

It was so warm . . . and getting warmer.[1]

The preceding narrative points out that there was a time, geologically not that long ago—about 10,000–12,000 years ago—when many parts of Earth were covered with massive sheets of ice. Moreover, the geologic record shows that this most recent ice-sheet-covering of large portions of Earth's surface is not a onetime phenomenon; instead, Earth has experienced several glaciation periods as well as interglacials like the one we are presently experiencing. Although the ice that Yurk and the cat experienced for the majority of their lives has now retreated from most of Europe, Asia, and North America, it has left traces of its influence across the whole face of the landscape in jagged mountain peaks, gouged-out upland valleys, swamps, changed river courses, and boulder-strewn, table-flat prairies in the lowlands.

Ice covers about 10 percent of all land and about 12 percent of the oceans. Most of this ice is contained in the polar ice sea, polar sheets and ice caps, valley glaciers, and piedmont glaciers formed by valley glaciers merging on a plain. In the grand scheme of geology of the present time, the glaciers of today are not that significant. However, it is the glaciation of the past, with its accompanying geologic evidence left behind by ancient glaciers that is important. This geologic record indicates that Earth's climate has undergone fluctuations in the past, and that the amount of Earth's surface covered by glaciers has been much larger in the past than in the present. In regards to the effects of past glaciation, one need only look at the topography of the western mountain ranges in the northern part of North America to view the significant depositional processes of glaciers.

# Glaciers

A **glacier** is a thick mass of slow-moving ice, consisting largely of recrystallized snow that shows evidence of downslope or outward movement due to the pull of gravity. Glaciers can only form at latitudes or elevations above the snowline (the elevation at which snow forms and remains present year round). Glaciers form in these areas if the snow becomes compacted, forcing out the air between the snowflakes. The weight of the overlying snow causes the snow to recrystallize and increase its grain size, until it increases its density and becomes a solid block of ice.

## TYPES OF GLACIERS

There are various types of glaciers including:

• **Mountain glaciers**—these are relatively small glaciers that occur at higher elevations in mountainous regions. A good example of mountain glaciers can be seen in the remaining glaciers of Glacier National Park, Montana (see figures 8.1 and 8.2). Note: The low snow/ice content of the cirque glaciers shown in the figures is due

**Figure 8.1.    Almost empty cirques (the bowl-shaped valleys at the head of a glacier). Glacier National Park, Montana. Photo by Frank R. Spellman.**

**Figure 8.2.    A more apparent and almost empty cirque. Elizabeth Lake, Glacier National Park, Montana. Photo by Frank R. Spellman.**

to Earth's recent warming trend and the warm weather patterns of the time of the year when photos were taken (July 2008).

- **Valley glaciers** are tongues of ice that accumulate and spill down a valley filling it with ice, perhaps for scores of miles. As cirque glaciers like the ones shown in figures 8.1 and 8.2 grow larger, they may flow down the valleys as valley glaciers. When a valley glacier extends down to sea level, it may carve a narrow valley into the coastline. These are called *fjord glaciers*, and the narrow valleys they carve and later become filled with seawater after the ice has melted are *fjords*. When a valley glacier extends down a valley and then covers a gentle slope beyond the mountain range, it is called a *piedmont glacier*. If valley glaciers cover a mountain range, they are called *ice caps*.
- **Continental glaciers (ice sheets)**—these are the largest glaciers. They cover Greenland and Antarctica and contain about 95 percent of all glacial ice on Earth.
- **Ice shelves**—these are sheets of ice floating on water and attached to land. They may extend hundreds of miles from land and reach thicknesses of several thousand feet.
- **Polar glaciers**—these are always below the melting point at the surface and do not produce any meltwater.
- **Temperate glaciers**—these are at a temperature and pressure level near the melting

point throughout the body except for a few feet of ice. This layer is subjected to annual temperature fluctuations.

## Glacier Characteristics

The primary characteristics displayed by glaciers are changes in size and movement. A glacier changes in size by the addition of snowfall, compaction, and recrystallization. This process is known as **accumulation**. Glaciers also shrink in size (due to temperature increases). This process is known as **ablation**.

Earth's gravity, pushing, pulling, and tugging almost everything toward Earth's surface is involved with the movement of glaciers. Gravity moves glaciers to lower elevations by two different processes:

- **Basal sliding**—this type of glacier movement occurs when a film of water at the base of the glacier reduces friction by lubricating the surface and allowing the whole glacier to slide across its underlying bed.
- **Internal flow**—called creep, forms fold structures and results from deformation of the ice crystal structure—the crystals slide over each other like playing cards in a deck. This type of flow is conducive to the formation of crevasses in the upper portions of the glacier. Generally, crevasses form when the lower portion of a glacier flows over sudden changes in topography.

## Did You Know?

Within a glacier the velocity constantly changes. The velocity is low next to the base of the glacier and where it is in contact with valley walls. The velocity increases toward the center and upper parts of the glacier.

## Glaciation

Glaciation is a geological process that modifies land surface by the action of glaciers. For those who study glaciation and glaciers, the fact that glaciations have occurred so recently in North America and Europe accounts for extant evidence, allowing the opportunity to study the undeniable results of glacial erosion and deposition. This is the case, of course, because the forces involved with erosion—weathering, mass wasting, and stream erosion—have not had enough time to remove the traces of glaciation from Earth's surface. Glaciated landscapes are the result of both glacial erosion (glaciers transport rocks and erode surfaces) and glacial deposition (glaciers transport material, then melt and deposit the material).

## GLACIAL EROSION

Glacial erosion has a powerful effect on land that has been buried by ice and has done much to shape our present world. Both valley and continental glaciers acquire tens of thousands of boulders and rock fragments, which, frozen into the sole of the glacier, act like thousands of abrasive files, gouging and rasping the rocks (and everything else) over which the glaciers pass. The rock surfaces display fluting, striation, and polishing effects of glacial erosion. The form and direction of these grooves show the direction in which the glaciers move.

Glacial erosion manifests itself in small-scale erosional features, landform production by mountain glaciers, and landforms produced by ice caps and ice sheets. These are described in the following.

**Small-scale erosional features**—these include glacial striations and polish. *Glacial striations* are long, parallel scratches and glacial grooves that are produced at the bottom of temperate glaciers by rocks embedded in the ice scraping against the rock underlying the glacier. *Glacial polish* is characteristic of rock that has a smooth surface produced as a result of fine-grained material embedded in the glacier acting like sandpaper on the underlying surface.

**Landforms produced by mountain glaciers**—these erosion-produced features include:

- **Cirques**—these are bowl-shaped valleys formed at the heads of glaciers and below arêtes and horned mountains; often they contain a small lake called a **tarn**.
- **Glacial Valleys**—these are valleys that once contained glacial ice and become eroded into a U shape in cross section. V-shaped valleys are the result of stream erosion.
- **Arêtes**—these are sharp ridges formed by headward glacial erosion.
- **Horns**—these are sharp, pyramidal mountain peaks formed when headward erosions of several glaciers intersect.
- **Hanging Valleys**—Yosemite's Bridal Veil Falls is a waterfall that plunges over a hanging valley. Generally, hanging valleys result in tributary streams that are not able to erode to the base level of the main stream; therefore, the tributary stream is left at higher elevation than the main stream, creating a hanging valley and sometimes spectacular waterfalls.
- **Fjords**—these are submerged, glacially deepened, narrow inlets with sheer, high sides, a U-shaped cross profile, and a submerged seaward sill largely formed of end moraine.

### Landforms produced by ice caps and ice sheets

- **Abrasional features**—these are small-scale abrasional features in the form of glacial polish and striations that occur in temperate environments beneath ice caps and ice sheets.
- **Streamlined forms**—sometimes called "basket of eggs" topography, the land

beneath a moving continental ice sheet is molded into smooth, elongated forms called *drumlins*. Drumlins are aligned in the direction of ice flow, their steeper, blunter ends point toward the direction from which the ice came.

## GLACIAL DEPOSITS

All sediment deposited as a result of glacial erosion is called **glacial drift**. It consists of rock fragments that are carried by the glacier on its surface, within the ice, and at its base.

### Ice land deposits

- **Till (or rock flour)**—this is nonsorted glacial drift deposited directly from ice. Consisting of a random mixture of different-sized fragments of angular rocks in a matrix of fine-grained, sand- to clay-sized fragments, till is produced by abrasion within the glacier. After undergoing diagenesis and turning to rock, till is called *tillite*.
- **Erratics**—this is a glacially deposited rock, fragment, or boulder that rests on a surface made of different rock. Erratics are often found miles from their source and by mapping the distribution pattern of erratics geologists can often determine the flow directions of the ice that carried them to their present locations. No one has described a glacial erratic better than William Wordsworth (1807) in his classic poem "The Leech-Gatherer (or, Resolution and Independence)."

> As a huge stone is sometimes seen to lie
> Couched on the bald top of an eminence;
> Wonder to all who do the same espy,
> By what means it could thither come, and whence;
> So that it seems a thing endued with sense:
> Like a Sea-beast crawled forth, that on a shelf
> Of rock or sand reposeth, there to sun itself.

- **Moraines**—these are mounds, ridges, or ground coverings of unsorted debris deposited by the melting away of a glacier. Depending on where it formed in relation to the glacier moraines can be:

  - *Ground moraines*—these are till-covered areas deposited beneath the glacier and result in a hummocky topography with lots of enclosed small basins.
  - *End moraines and terminal moraines*—these are ridges of unconsolidated debris deposited at the low elevation end of a glacier as the ice retreats due to ablation (melting). They usually reflect the shape of the glacier's terminus.
  - *Lateral moraines*—these are till deposits that were deposited along the sides of mountain glaciers.
  - *Medial moraines*—when two valley glaciers meet to form a larger glacier, the rock debris along the sides of both glaciers merge to form a medial moraine (runs down the center of a valley floor).

- *Glacial marine drift (icebergs)*—these are glaciers that reach lake shores or oceans and calve off into large icebergs that then float on the water surface until they melt. The rock debris that the icebergs contain is deposited on the lakebed or ocean floor when the iceberg melts.

**Stratified drift**—this is glacial drift that can be picked up and moved by meltwater streams, which can then deposit the material as stratified drift.

- **Outwash Plains**—melt runoff at the end of a glacier is usually choked with sediment and forms braided streams, which deposit poorly sorted stratified sediment in an outwash plain—they usually are flat, interlocking alluvial fans.
- **Outwash Terraces**—these form if the outwash streams cut down into their outwash deposits, forming river terraces called outwash terraces.
- **Kettle Holes**—these are depressions (sometimes filled by lakes; e.g., Minnesota, the land of a thousand lakes) due to melting of large blocks of stagnant ice, found in any typical glacial deposit.
- **Kames**—these are isolated hills of stratified material formed from debris that fell into openings in retreating or stagnant ice.
- **Eskers**—these are long, narrow, and often branching sinuous ridges of poorly sorted gravel and sand formed by deposition from former glacier streams.

# Glacial Ages

Before beginning a discussion of the past, it is important to define the era referred to when we say "the past." Tables 8.1 and 8.2 are provided to assist us in making this definition. Table 8.1 gives the entire expanse of time from Earth's beginning to present. Table 8.2 provides the sequence of geological epochs over the past 65 million years, as dated by modern methods. The Paleocene through Pliocene together make up the Tertiary Period; the Pleistocene and the Holocene compose the Quaternary Period.

When we think about the prehistoric past, two things generally come to mind—ice ages and dinosaurs. Of course, in the immense span of time prehistory covers, those two eras represent only a brief moment in time, so let's look at what we know about the past and about Earth's climate and conditions. One thing to consider—geological history shows us that the normal climate of Earth was so warm that subtropical weather reached to latitudes of 60°N and S, and polar ice was entirely absent.

Only during less than about 1 percent of Earth's history did glaciers advance and reach as far south as what is now the temperate zone of the northern hemisphere. The latest such advance, which began about 1,000,000 years ago, was marked by geological upheaval and (perhaps) the advent of human life on earth. During this time, vast ice sheets advanced and retreated, grinding their way over the continents.

## A TIME OF ICE

Nearly 2 billion years ago, the oldest known glacial epoch occurred. A series of deposits of glacial origin in southern Canada, extending east to west about 1,000 miles,

**Table 8.1.  Geologic Eras and Periods**

| ERA | Period | Millions of Years Before Present |
|---|---|---|
| Cenozoic | Quaternary | 2.5–present |
| | Tertiary | 65–2.5 |
| | | |
| Mesozoic | Cretaceous | 135–65 |
| | Jurassic | 190–135 |
| | Triassic | 225–190 |
| | | |
| Paleozoic | Permian | 280–225 |
| | Pennsylvanian | 320–280 |
| | Mississippian | 345–320 |
| | Devonian | 400–345 |
| | Silurian | 440–400 |
| | Ordovician | 500–440 |
| | Cambrian | 570–500 |
| | | |
| Precambrian | | 4,600–570 |

**Table 8.2.  Epochs of Geologic Time**

| Epoch | Million Years Ago |
|---|---|
| Holocene | 01–0 |
| Pleistocene | 1.6–.01 |
| Pliocene | 5–1.6 |
| Miocene | 24–5 |
| Oligocene | 35–24 |
| Eocene | 58–35 |
| Paleocene | 65–58 |

shows us that within the last billion years or so, apparently at least six major phases of massive, significant climatic cooling and consequent glaciation occurred at intervals of about 150 million years. Each lasted perhaps as long as 50 million years.

Examination of core samples of land and oceanic sediment clearly indicate that in more recent times many alternating episodes of warmer and colder conditions occurred over the last 2 million years (during the middle and early Pleistocene epochs). In the last million years, at least eight such cycles have occurred, with the warm part of the cycle lasting a relatively short interval.

During the Great Ice Age (the Pleistocene epoch), ice advances began—a series of them that at times covered over a quarter of the earth's land surface. Great sheets of ice thousands of feet thick, glaciers moved across North America again and again, reaching as far south as the Great Lakes. An ice sheet thousands of feet thick spread over Northern Europe, sculpting the land and leaving behind lakes, swamps, and terminal moraines as far south as Switzerland. Each succeeding glacial advance was apparently more severe than the previous one. Evidence indicates that the most severe began about 50,000 years ago and ended about 10,000 years ago. Several interglacial stages separated the glacial advances, melting the ice. Average temperatures were higher than ours today.

Temperatures were higher than today? Yes. Think about that as we proceed.

Because one-tenth of the globe's surface is still covered by glacial ice, scientists consider the earth still to be in a glacial stage. The ice sheet has been retreating since the climax of the last glacial advance, and world climates, although fluctuating, are slowly warming.

From observations and from well-kept records, we know that the ice sheet is in a retreating stage. The records clearly show that a marked worldwide retreat of ice has occurred over the last hundred years. World famous for its fifty glaciers and 200 lakes, Glacier National Park in Montana does not present the same visual experiences it did a hundred years ago. A ten-foot pole put into place at the terminal edge of one of the main glaciers holds a "1939" sign. The sign, still in place, is now several hundred feet from the terminal end of the glacier, which has retreated back up the slope of the mountain. Swiss resorts built during the early 1900s to offer scenic glacial views now have no ice in sight. Theoretically, if glacial retreat continues, melting all of the world's ice supply, sea levels would rise more than 200 feet, flooding many of the world's major cities. New York and Boston would become aquariums.

The question of what causes ice ages is one that scientists still grapple with. Theories range from changing ocean currents to sunspot cycles. On one fact we are absolutely certain, however. An ice age event occurs because of a change in Earth's climate.

But what could cause such a drastic change?

Climate results from uneven heat distribution over Earth's surface. It's caused by Earth's tilt—the angle between Earth's orbital plane around the sun and its rotational axis. This angle is currently 23.5 degrees.

The angle has changed. It has not always been 23.5 degrees. The angle, of course, affects the amount of solar energy that reaches Earth, and where it falls.

The heat balance of the earth, which is driven mostly by the concentration of carbon dioxide ($CO_2$) in the atmosphere, also affects long-term climate.

If the pattern of solar radiation changes and/or if the amount of $CO_2$ changes, climate change can result. Abundant evidence that the earth does undergo climatic change exists, and we know that climatic change can be a limiting factor for the evolution of many species.

Evidence (primarily from soil core samples and topographical formations) tells us that changes in climate include events such as periodic ice ages, characterized by glacial and interglacial periods. Long glacial periods have lasted up to 100,000 years, where temperatures decreased about 9°F and ice covered most of the planet. Short periods lasted up to 12,000 years, with temperatures decreasing by 5°F, and ice covering latitudes above 40 N. Smaller periods (the "Little Ice Age," which occurred from about AD 1000–1850) had about a 3°F drop in temperature. Note: Despite its name, "Little Ice Age" was a time of severe winters and violent storms, not a true glacial period.

These ages may or may not be significant—but consider that we are presently in an interglacial period, and that we may be reaching its apogee. What does that mean? No one knows with any certainty. Let's look at the effects of ice ages—the effects we think we know about.

Changes in sea levels could occur. Sea level could drop by about 100 meters in a full-blown ice age, exposing the continental shelves. Increased deposition during melt would change the composition of the exposed continental shelves. Less evaporation would change the hydrological cycle. Significant landscape changes could occur—on the scale of the Great Lakes formation. Drainage patterns and topsoil characteristics throughout most of the world would change. Flooding on a massive scale could occur.

How would these changes affect us? That depends on whether you live in Northern Europe, Canada, Seattle, Washington, around the Great Lakes, or near a seashore.

We are not sure what causes ice ages, but we have some theories (don't people always have theories?). To generate a full-blown (massive ice sheet covering most of the globe) ice age, scientists point out that certain periodic or cyclic events or happenings must occur. Periodic fluctuations would have to affect the solar cycle, for instance. However, we have no definitive proof that this has ever occurred.

Another theory speculates periods of volcanic activity could generate masses of volcanic dust that would block or filter heat from the sun. This would cool down the earth. Some speculate that the carbon dioxide cycle would have to be periodic/cyclic to bring about periods of climate change. There is reference to a so-called factor-2 reduction, causing a 7°F temperature drop worldwide. Others speculate that another global ice age could be brought about by increased precipitation at the poles due to changing orientation of continental land masses. Others theorize that a global ice age would result if changes in the mean temperatures of ocean currents occurred. But the question is how? By what mechanism?

Are these plausible theories? No one is sure—this is speculation only.

Speculation aside, what are the most probable causes of ice ages on Earth? According to the Milankovitch hypothesis, ice age occurrences are governed by a combination of factors: (1) Earth's change of altitude in relation to the sun (the way it tilts in a 41,000-year cycle and at the same time wobbles on its axis in a 22,000-year cycle) making the time of its closest approach to the sun come at different seasons; and (2) the 92,000-year cycle of eccentricity in its orbit round the sun, changing it from an elliptical to a near circular orbit, the severest period of an ice age coinciding with the approach to circularity.

So what does all this mean? We have a lot of speculation about ice ages and their causes and their effects. This is the bottom line: We know that ice ages occurred—we know that they caused certain things to occur (formation of the Great Lakes, etc.), and while there is a lot we do not know, we recognize the possibility of recurrent ice ages.

Lots of possibilities exist. Right now, no single theory is sound, and doubtless many factors are involved. Keep in mind that the possibility does exist that we are still in the Pleistocene Ice Age. It may reach another maximum in another 60,000 years or so.

## WARM WINTER

Earlier, when we discussed possible causes of glaciation and subsequent climatic cooling, we were left hanging without adequate explanation. This is what I call the real inconvenient truth: We simply don't know what we don't know. In this section, we discuss how we know what we think we do know about climatic change.

The headlines we see in the paper sound authoritative: "1997 Was the Warmest Year on Record." "Scientists Discover Ozone Hole Is Larger Than Ever." "Record Quantities of Carbon Dioxide Detected in Atmosphere." Or maybe you saw the one that read: "January 1998 Was the Third Warmest January on Record." Other reports indicate we are undergoing a warming trend. But conflicting reports abound. What do we know about climate change?

Two environmentally significant events took place late in 1997: El Niño's return and the Kyoto Conference: Summit on Global Warming and Climate Change. News reports blamed El Niño—the coupled ocean-atmosphere phenomenon of the eastern and western Pacific Ocean—for just about everything and anything that had to do with weather conditions throughout the world. Some occurrences were indeed El Niño-related or generated: the out-of-control fires, droughts, floods, and the stretches of dead coral, no sign of fish in the water and few birds around certain Pacific atolls. The devastating storms that struck the west coasts of South America, Mexico and California were also probably El Niño related. El Niño's affect on the 1997 hurricane season, one of the mildest on record, is not in question, either.

But does a connection exist between El Niño and global warming or global climate change?

On December 7, 1997, the Associated Press reported that while delegates at the global climate conference in Kyoto haggled over greenhouse gases and emission limits, a compelling question has emerged, "Is global warming fueling El Niño?" NOAA (2008) stated emphatically that El Ninos are not caused by global warming. Clear evidence exists that El Ninos have been present for thousands, and some indication suggests maybe millions, of years. One thing seems certain; nobody knows for sure.

Why aren't we sure? Because we need more information than we have today. Our amount of recorded data is paltry; what information we have suggests, however, that El Niño is getting stronger and more frequent.

Some scientists fear that El Niño's increasing frequency and intensity (records show that two of this century's three worst El Niños came in 1982 and 1997) may be linked to global warming. At the Kyoto Conference, experts said the hotter atmosphere is heating up the world's oceans, setting the stage for more frequent and extreme El Niños.

Weather-related phenomena seem to be intensifying throughout the globe. Can we be sure that this is related to global warming yet? No. Without more data—more time—more science, we cannot be sure.

Should we be concerned? Yes. According to the Associated Press coverage of the Kyoto Conference, scientist Richard Fairbanks reported that he found startling evidence of our need for concern. During two months of scientific experiments on

Christmas Island (the world's largest atoll in the Pacific Ocean) conducted in autumn 1997, he discovered a frightening scene. The water surrounding the atoll was 7°F higher than average for that time of year, which upset the balance of the environmental system. According to Fairbanks, 40 percent of the coral was dead, the warmer water had killed off or driven away fish, and the atoll's plentiful bird population was almost completely gone.

El Niños have an acute impact on the globe; that is not in doubt. However, we do not know if it is caused by or intensified because of global warming. What do we know about global warming and climate change?

*USA Today* (December 1997) discussed the results of a report issued by the Intergovernmental Panel on Climate Change. They interviewed Jerry Mahlman of the National Oceanic and Atmospheric Administration and Princeton University, and presented the following information about what most scientists agree on:

- There is a natural "greenhouse effect," and scientists know how it works—and that without it, Earth would freeze.
- The Earth undergoes normal cycles or warming and cooling on grand scales. Ice ages occur every 20,000 to 100,000 years.
- Globally, average temperatures have risen 1°F in the past 100 years, within the range that might occur normally.
- The level of man-made carbon dioxide in the atmosphere has risen 30 percent since the beginning of the Industrial Revolution in the nineteenth century, and is still rising.
- Levels of man-made carbon dioxide will double in the atmosphere over the next 100 years, generating a rise in global average temperatures of about 3.5°F (larger than the natural swings in temperature that have occurred over the past 10,000 years).
- By 2050, temperatures will rise much higher in northern latitudes than the increase in global average temperatures. Substantial amounts of northern sea ice will melt, and snow and rain in the northern hemisphere will increase.
- As the climate warms, the rate of evaporation will rise, further increasing warming. Water vapor also reflects heat back to Earth.

## WHAT WE THINK WE KNOW ABOUT GLOBAL WARMING

What is global warming? To answer this question we need to discuss the "greenhouse effect." As mentioned, water vapor, carbon dioxide, and other atmospheric gases (greenhouse gases) help warm the earth. How does greenhouse effect actually work? Let's take a closer look at this phenomenon.

### Greenhouse Effect

Earth's greenhouse effect, of course, took its name because of its similarity to how a greenhouse's glass walls and ceilings are largely transparent to short-wave radiation from the sun, causing surfaces and objects inside the greenhouse to absorb the radiation. The radiation, once absorbed, transforms into long-wave (infrared) radiation (heat), and radiates back from the greenhouse interior. But the glass prevents the long-wave radiation from escaping again, absorbing the warm rays. The interior of the greenhouse becomes much warmer than the air outside, because of the heat trapped inside.

Earth and its atmosphere undergo a process very similar to this—the greenhouse effect. Short-wave and visible radiation reaching Earth is absorbed by the surface as heat. The long heat waves radiate back out toward space, but the atmosphere absorbs many of them, trapping them. This natural and balanced process is essential to supporting our life systems on earth. Changes in the atmosphere can radically change the amount of absorption (therefore the amount of heat) the earth's atmosphere retains. In recent decades, scientists speculate that various air pollutants have caused the atmosphere to absorb more heat. At the local level with air pollution, the greenhouse effect causes heat islands in and around urban centers, a widely recognized phenomenon.

The main contributors to this effect are the greenhouse gases: water vapor, carbon dioxide, carbon monoxide, methane, volatile organic compounds (VOCs), nitrogen oxides, chlorofluorocarbons (CFCs), and surface ozone. These gases cause a general climatic warming by delaying the escape of infrared radiation from the earth into space. Scientists stress this is a natural process—indeed, if the "normal" greenhouse effect did not exist, the earth would be 33°C cooler than it presently is (Hansen et al., 1986).

Human activities are now rapidly intensifying the natural phenomenon of Earth's greenhouse effect, which may lead to problems of warming on a global scale. Much debate, confusion, and speculation about this potential consequence is underway because scientists cannot yet agree about whether the recently perceived worldwide warming trend is because of greenhouse gases, due to some other cause, or simply a wider variation in the normal heating and cooling trends they have been studying. Unchecked, the greenhouse effect may lead to significant global warming, with profound effects upon our lives and our environment. Human impact on the greenhouse effect is real; it has been measured and detected. The rate at which the greenhouse effect is intensifying is now more than five times what it was during the last century (Hansen et al., 1989).

Supporters of the global warming theory base their assumptions on man's altering of the earth's normal and necessary greenhouse effect. The human activities they blame for increases of greenhouse gases include burning of fossil fuels, deforestation, and use of certain aerosols and refrigerants. These gases have increased how much heat remains trapped in the earth's atmosphere, gradually increasing the temperature of the whole globe.

From information based on recent or short-term observation, many scientists note that the last decade has been the warmest since temperature recordings began in the late nineteenth century. They see that the general rise in temperature over the last century coincides with the Industrial Revolution and its accompanying increase in fossil fuel use. Other evidence supports the global warming theory. In places that are synonymous with ice and snow—the Arctic and Antarctica, for example, we see evidence of receding ice and snow cover.

Trying to pin down definitively whether or not changing our anthropogenic activities could have any significant effect on lessening global warming, though, is difficult. Scientists look at temperature variations over thousands and even millions of years, taking a long-term view of Earth's climate. The variations in Earth's climate are wide enough that they cannot definitively show that global warming is anything more than another short-term variation. Historical records that have shown the Earth's temperature does vary widely, growing colder with ice ages and then warming again, and because we cannot be certain of the causes of those climate changes, we cannot be certain of what appears to be the current warming trend.

Still, debate abounds for the argument that our climate is warming and our activities are part of the equation. The 1980s saw nine of the twelve warmest temperatures ever recorded, and the Earth's average surface temperature has risen approximately 0.6°C (1°F) in the last century (USEPA, 1995). *Time* magazine (1998) reports that scientists are increasingly convinced that because of the buildup in the atmosphere of carbon dioxide and other gases produced in large part by the burning of fossil fuels, the earth is getting hotter. Each month from January through July 1998, for example, set a new average global temperature record, and if that trend continued, the surface temperature of the earth could rise by about 1.8 to 6.3°F by 2100. At the same time, others offer as evidence that the 1980s also saw three of the coldest years: 1984, 1985, and 1986.

What is really going on? We cannot be certain. Assuming that we are indeed seeing long-term global warming, we must determine what causes it. But again, we face the problem that scientists cannot be sure of the greenhouse effect's precise causes. Our current possible trend in global warming may simply be part of a much longer trend of warming since the last ice age. We have learned much in the past two centuries of science, but little is actually known about the causes of the worldwide global cooling and warming that sent the earth through major and smaller ice ages. The data we need reaches back over millennia. We simply do not possess enough long-term data to support our theories.

Currently, scientists can point to six factors they think could be involved in long-term global warming and cooling.

1. Long-term global warming and cooling could result if changes in the earth's position relative to the sun occur (i.e., the earth's orbit around the sun), with higher temperatures when the two are closer together and lower when further apart.
2. Long-term global warming and cooling could result if major catastrophes (meteor impacts or massive volcanic eruptions) that throw pollutants into the atmosphere that can block out solar radiation occur.
3. Long-term global warming and cooling could result if changes in albedo (reflectivity of earth's surface) occur. If the earth's surface were more reflective, for example, the amount of solar radiation radiated back toward space instead of absorbed would increase, lowering temperatures on earth.

4. Long-term global warming and cooling could result if the amount of radiation emitted by the sun changes.

5. Long-term global warming and cooling could result if the shape and relationship of the land and oceans change.

6. Long-term global warming and cooling could result if the composition of the atmosphere changes.

*If the composition of the atmosphere changes*—this final factor, of course, defines our present concern: Have human activities had a cumulative impact large enough to affect the total temperature and climate of earth? Right now, we cannot be sure. The problem concerns us, and we are alert to it, but we are not certain.

## THE BOTTOM LINE

When news media personnel, would-be presidential candidates, and doomsayers in general make their dire warnings about global climate change (specifically that the earth is getting warmer), keep in mind the opening fictional segment to this chapter. Recall that Yurk's and the cat's lives ended during the transition from ice age to an interglacial age. Two points should be clear from this: (1) The transition from ice age to interglacial (a warm period) is well documented with considerable actual evidence to back it up; (2) keep in mind that when Yurk and the cat were experiencing the transition from ice age to interglacial they certainly would not have been able to point the finger of blame for the warming trend on man and/or his actions. Simply put, the fact is, at that time, man made little contribution of carbon dioxide, CFCs, or any other chemical substance to earth's atmosphere. Yet, the earth warmed up anyway.

# Chapter Review Questions

8.1 _____ glaciers are relatively small glaciers that occur at higher elevations in mountainous regions.

8.2 _____ glaciers extend down to sea level where they may carve a narrow valley into the coastline.

8.3 _____ type of glacier movement occurs when a film of water at the base of the glacier reduces friction by lubricating the surface and allowing the whole glacier to slide across its underlying bed.

8.4 _____ is a bowl-shaped valley formed at the heads of glaciers.

8.5 _____ is a small cirque lake.

8.6 _____ are sharp ridges formed by headward glacial erosion.

8.7 _____ are streamlined, elongated forms.

8.8 Rock flour: _____.

8.9 Glacially deposited rock, fragment, or boulder that rests on a surface made of different rock: _____.

8.10 _____ are long, narrow, and often branching sinuous ridges of poorly sorted gravel and sand formed by deposition from former glacier streams.

# References and Recommended Reading

Associated Press. 1997. Does warming feed El Niño? *Virginian-Pilot* (Norfolk), December 7, p. A-15.

Associated Press. 1998. Ozone hole over Antarctica at record size. *Lancaster New Era* (Lancaster, PA), September 28.

Associated Press. 1998. Tougher air pollution standards too costly, Midwestern states say. *Lancaster New Era* (Lancaster, PA), September 25.

Chernicoff, S. 1999. *Geology.* Boston: Houghton Mifflin

Dolan, E. F. 1991. *Our Poisoned Sky.* New York: Cobblehill Book.

Hansen, J. E., et al. 1986. Climate sensitivity to increasing greenhouse gases. In *Greenhouse Effect and Sea Level Rise: A Challenge for This Generation*, ed. M. C. Barth and J. G. Titus. New York: Van Nostrand Reinhold.

Hansen, J. E., et al. 1989. Greenhouse effect of chlorofluorocarbons and other trace gases. *Journal of Geophysical Research* 94 (November) pp. 16, 417–421.

NOAA. 2008. Global Warming: Frequently asked questions. Accessed 11/21/08 at http://lwf.ncdc.noaa.gov/oa/climate/globalwarming.html.

Tarbuck, E. J., and Lutgens, F. K. 2000. *Earth Science.* Upper Saddle River, NJ: Prentice Hall.

*Time.* 1998. Global Warming: It's Here . . . And Almost Certain to Get Worse. *Time*, August 24.

*USA Today.* 1997. "Global warming: Politics and economics further complicate the issue." December 1, pp. A-1, 2.

# Notes

1. This account of Yurk and Smilodon is a case study devised and used by the author in his 400/500-level Industrial Environmental Management course at Old Dominion University. Students are required to make their own interpretation of the author's message (if there is one?). Interpretations vary greatly, but most students get it: both Yurk and Smilodon (at the end of their lives) experienced the transition from Earth's last glacial period to the present interglacial warming period. Further, they understand that 12,000 years ago there were no humankind contributions to the global climate change—it was a natural, cyclical phenomenon. F. R. Spellman, *The Science of Environmental Pollution* (Lancaster, PA: Technomic Publishing Company, 1998).

# CHAPTER 9

# Wind Erosion and Mass Wasting

During a recent research outing to several national parks in the western United States, I stopped at several locations and photographed various natural wonders. One of the focal points of study was on the weathering processes that are discussed in this chapter. The natural bridges, such as the one shown in figure 9.1 and the natural arches shown in figures 9.2–9.5, all are a result of some form of weathering; thus, they are highlighted in this chapter.

There was a time not that long ago when many believed that the main difference between natural bridges and natural arches was that the natural bridges were formed by water erosion and natural arches were formed by wind erosion. Contrary to popular belief or myth, however, wind is not a significant factor in the formation of natural arches or other natural formations. Substantial studies have shown that natural arches and natural bridges are formed by many different processes of erosion that contribute to the natural, selective removal of rock. Every process relevant to natural arch formation involves the action of water, gravity, temperature variation or, tectonic pressure on rock.

**Figure 9.1. Natural Bridge. Natural Bridge, Virginia. Photo by F. R. Spellman.**

**Figure 9.2.    Natural Arch. Monument Valley, Utah. Photo by F. R. Spellman**

Again, wind is not a significant agent in natural arch formation. Wind does act to disperse the loose grains that result from microscopic erosion. Moreover, sandstorms can scour and polish already existing arches. The point to remember is that wind never alone creates them (Barnes, 1987; Vreeland, 1994).

# Wind Erosion

As prefaced above, wind action or erosion is very limited in extent and effect. It is largely confined to desert regions, but even there it is limited to a height of about 18 inches above ground level. Wind does have the power, however, to transport, deposit, and erode sediment. In this chapter we will discuss each of the aspects of the wind because they are important in any study of geology.

## Did You Know?

Wind is common in deserts because the air near the surface is heated and rises, cooler air comes in to replace hot rising air, and this movement of air results in winds. Also, arid desert regions have little or no soil moisture to hold rock and mineral fragments.

**Figure 9.3.   Eye-to-the-Sky Natural Arch. Monument Valley, Utah. Photo by F. R. Spellman**

## WIND SEDIMENT TRANSPORT

Sediment near the ground surface is transported by wind in a process called *saltation* (from Latin, *saltus*, "leap"). Similar to what occurs in the bed load of streams, wind saltation refers to short jumps (leaps) of grains dislodged from the surface and leaping a short distance. As the grains fall back to the surface they dislodge other grains that then get carried by wind until they collide with the ground to dislodge other particles. Above ground level, wind can swoop down to the surface and lift smaller particles, suspending them (making them airborne), and may carry them long distances (see figure 9.6).

## Did You Know?

Sand ripples occur as a result of large grains accumulating as smaller grains are transported away. Ripples form in lines perpendicular to wind direction. Wind-blown dust in sand-sized particles generally does not travel very far in the wind, but smaller-sized fragments can be suspended in the wind for much larger distances.

**Figure 9.4.    Delicate Arch. Arches National Park, Moab, Utah. Photo by F. R. Spellman**

**Wind-Driven Erosion**

As mentioned, wind by itself has little if any effect on solid rock. But in arid and semiarid regions, wind can be an effective geologic agent anywhere that it is strong enough (possesses high enough velocity) to pick up a load of rock fragments that may become effective tools of erosion in the land-forming process. Wind can erode by **deflation** and **abrasion**.

*Deflation*

The process of deflation (or blowing away) is the lowering of the land surface due to removal of fine-grained particles by the wind. Deflation removes the finer-grained particles at the surface, eventually resulting in a relatively smooth surface composed only of the coarser grained fragments that cannot be transported by the wind. Such a coarse-grained surface is called *desert pavement* (see figure 9. 7). Some of these coarser-grained fragments may exhibit a dark, enamel-like coat of iron or manganese called *desert varnish*.

Deflation may create several types of distinctive features. For example, *lag gravels* are formed when the wind blows away finer rock particles, leaving behind a residue of coarse gravel and stones. *Blowouts* may be developed where wind has scooped out soft unconsolidated rocks and soil.

**Figure 9.5.   North and South Windows. Arches National Park, Moab, Utah. Photo by F. R. Spellman.**

### Abrasion

The wind abrades (sandblasts) by picking up sand and dust particles that are transported as part of its load. Abrasion is restricted to a distance of about a meter or two above the surface because sand grains are lifted a short distance. The destructive action of these windblown abrasives may wear away wooden telephone poles and fence posts, and abrade, scour, or groove solid rock surfaces.

Wind abrasion also plays a part in the development of such landforms as isolated rocks (pedestals and table rocks), which have had their bases undercut by wind-blown sand and grit (see figure 9.8). *Ventifacts* are another interesting and relatively common product of wind erosion. These are any bedrock surface or stone or pebble that has been abraded or shaped by wind-blown sediment in a process similar to sandblasting. Ventifacts are formed when the wind blows sand against the side of the stone, shaping it into a flat, polished surface. On a much larger scale, elongate ridges called *yardangs* form by the abrasion and streamlining of rock structures oriented parallel to the prevailing wind direction.

## WIND DEPOSITION

The velocity of the wind and the size, shape, and weight of the rock particles determines the manner in which wind carries its load. Wind-transported materials are most

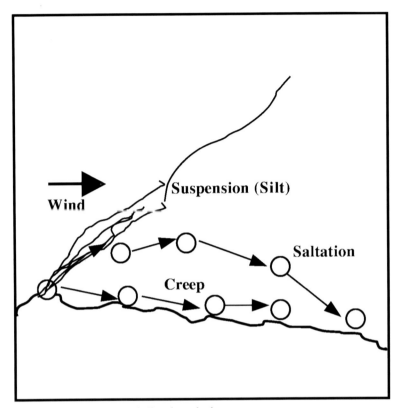

**Figure 9.6.    Sediment transportation by wind.**

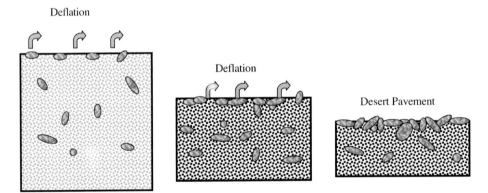

**Figure 9.7.    Wind-driven deflation processes.**

Figure 9.8.    Seeming to defy gravity, Balanced Rock (Arches National Park, Moab, Utah) has a harder cap rock that somewhat protects the more easily eroded base; eventually the double hammering of water and wind erosion will cause it to disintegrate, leaving a pile of rocky debris as a reminder of the power of erosion. Photo by Frank R. Spellman.

commonly derived from flood plains, beach sands, glacial deposits, volcanic explosions, and dried lake bottoms—places containing light ash and loose, weathered rock fragments.

The wind is capable of transporting large quantities of material for very great distances. The wind deposits sediment when its velocity decreases to the point where the particles can no longer be transported. Initially (in a strong wind) part of the sediment load rolls or slides along the ground (bed load). Some sand particles move by a series of leaping or bounding movements (saltation). And lighter dust may be transported upward (suspension) into higher, faster-moving wind currents, traveling many thousands of miles.

As mentioned, the wind will begin to deposit its load when its velocity is decreased, or when the air is washed clean by falling rain or snow. A decrease in wind velocity may also be brought about when the wind strikes some barrier-type obstacle (fences, trees, rocks, human made structures) in its path. As the air moves over the top of the obstacle, streamlines converge and the velocity increases. After passing over the obstacle, the streamlines diverge and the velocity decreases. As the velocity decreases, some of the load can no longer be held in suspension and thus drops out to form a deposit. The major types of wind-blown or eolian deposits are **dunes** and **loess**.

*Dunes*

Sand dunes are asymmetrical mounds with a gentle slope in the upwind direction and steep slope on the downwind side (see figure 9.9). Dunes vary greatly in size and shape and form when there is a ready supply of sand, a steady wind, and some kind of obstacle or barrier such as rocks, fences, or vegetation to trap some of the sand. Sand dunes form when moving air slows down on the downwind side of an obstacle (see figure 9.9). Dunes may reach heights up to 500 meters and cover large areas. Types of sand dunes include barchan, transverse, longitudinal, and parabolic.

• **Barchan dunes**—these are crescent-shaped dunes characterized by two long, curved extensions pointing in the direction of the wind, and a curved slip face on the downwind side of the dune (see figure 9.10[a]). These dunes are formed in areas where winds blow steadily and from a single direction.
• **Transverse dunes**—these form along sea costs and lakeshores and may be fifteen feet high and half a mile in length. Transverse dunes develop with their long axis at right angles to the wind (see figure 9.10[b]).
• **Longitudinal dunes**—these are long, ridgelike dunes that develop parallel to the wind (see figure 9.10[c]).
• **Parabolic dunes**—these are U-shaped dunes with an open end facing upwind. They are usually stabilized by vegetation and occur where there is abundant vegetation, a constant wind direction, and an abundant sand supply (see figure 9.10[d]).

## LOESS

Loess is a yellowish, fine-grained, nonstratified material carried by the wind and accumulated in deposits of dust. The materials forming loess are derived from surface dust originating primarily in deserts, river flood plains, deltas, and glacial outwash deposits. Loess is cohesive and possesses the property of forming steep bluffs with vertical faces such as the deposits found in the pampas of Argentina, and the lower Mississippi River Valley.

# Mass Wasting

Mass wasting, or mass movement, takes place as earth materials (loose uncemented mixture of soil and rock particles known as *regolith*) move downslope in response to

**Figure 9.9.   Profile of typical sand dune. Arrows denote paths of wind currents.**

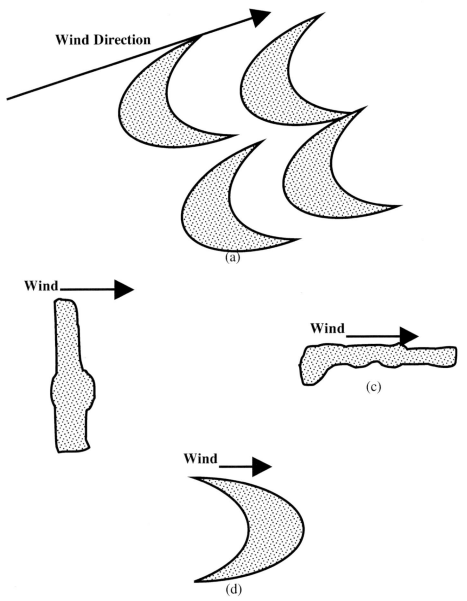

**Figure 9.10.** (a) Barchan dune; (b) transverse dunes; (c) longitudinal dune; (d) parabolic dune.

gravity without the aid of a transporting medium such as water, ice, or wind—though these factors play a role in regolith movement. This type of erosion is apt to occur in any area with slopes steep enough to allow downward movement of rock debris. Some of the factors that help gravity in mass wasting are discussed below.

## GRAVITY

The heavy hand of gravity constantly pulls everything, everywhere toward Earth's surface. On a flat surface parallel to Earth's surface, the constant force of gravity acts downward. This downward force prevents gravitational movement of any material that remains on such a flat surface.

On a slope, the force of gravity can be resolved into two components: a component acting perpendicular to the slope, and a component acting tangential to the slope. Thus, material on a slope is pulled inward in a direction that is perpendicular to the slope (the glue) (see figure 9.11[a]). This helps prevent material from sliding downward. However, as stated previously, on a slope, another component of gravity exerts a force (a constant tug) that acts to pull material down, parallel to the surface of the slope. Known as *shear stress*, this force of gravity exerts stress in direct relationship to the steepness of the slope. That is, shear stress increases as the slope steepens. In response to increased shear stress, the perpendicular force (the glue) of gravity decreases (see figure 9.11[b]).

## Did You Know?

When shear on a slope decreases, material may still be stuck to the slope and be prevented from moving downward by the force of friction. It may be held in place by the frictional contact between the particles making up that material. Contact between the surfaces of the particles creates a certain amount of tension that holds the particles in place at an angle. The steepest angle at which loose material on a slope remains motionless is called the angle of repose (generally about thirty-five degrees). Particles with angled edges that catch on each other also tend to have a higher angle of repose than those that have become rounded through weathering and that simply roll over each other.

## WATER

Even though mass wasting may occur in either wet or dry materials, water greatly facilitates downslope movements; it is an important agent in the process of mass wasting. Water will either help hold material together (act as glue—demonstrated in building beach sandcastles with slightly dampened sand), increasing its angle of repose, or cause it to slide downward like a liquid (acting like a lubricant). Water may soften clays, making them slippery, add weight to the rock mass, and, in large amounts, may actually force rock particles apart, thus reducing soil cohesion.

## FREEZING AND THAWING

Earlier the erosive power of frost wedging (water contained in rock and soil expands when frozen) was pointed out. Mass wasting in cold climates is governed by the fact

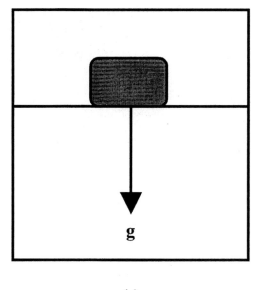

(a)

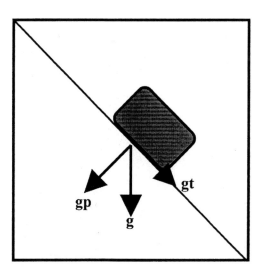

(b)

Figure 9.11. (a) Gravity acting perpendicular to the surface. (b) The perpendicular component (the glue) of gravity, *gp*, helps to hold the material in place on the slope.

that water is frozen as ice during long periods of the year, especially in high altitude regions. Ice, although it is solid, does have the ability to flow (glacial-movement effect), and alternate periods of freezing and thawing can also contribute to movement; in some instances ice expansion may be great enough to force rocks downhill.

## UNDERCUTTING

Undercutting occurs when streams erode their banks or surf action along a coast undercuts a slope, making it unstable. Undercutting can also occur when human-made excavations remove support and allow overlying material to fall.

## ORGANIC ACTIVITIES

Whenever animals burrow into the ground, they disturb soil materials, casting rocks out of their holes as they dig; these are commonly piled up downslope. Eventually, weather conditions and the constant force applied by gravity can put these piles into motion. Animals also contribute to mass wasting whenever they walk on soil surfaces; their motions can knock materials downhill.

## SHOCK WAVES OR VIBRATIONS

A sudden strong shock or vibration, such as an earthquake, faulting, blasting, and heavy traffic can trigger slope instability. Minor shocks, like heavy vehicles rambling down the road, trees blowing in the wind, or human-made explosions can also trigger mass-wasting events such as landslides.

# Kinds of Mass Movements

A landslide is a mass movement that occurs suddenly and violently. In contrast, soil creep is mass movement that is almost imperceptible. These processes can be divided into two broad categories: rapid and slow movements. *Rapid movements* include landslides, slumps, mudflows, and earthflows. *Slow movements* include soil creep and solifluction.

### Rapid Movements

- **Landslides**—these are by far the most spectacular and violent of all mass movements. Landslides are characterized by the sudden movement of great quantities of rock and soil downslope. Such movements typically occur on steep slopes that have large accumulations of weathered material. Precipitation in the form of rain or snow may seep into the mass of steeply sloping rock debris, adding sufficient weight to start the entire mass sliding.

- **Slumps**—these special landslides occur along curved surfaces. The upper surface of each slump block remains relatively undisturbed, as do the individual blocks. Slumps leave arcuate (L. "curved like a bow") scars or depressions on the hill slope. Heavy rains or earthquakes usually trigger slumps. A slump is a common occurrence along the banks of streams or the walls of steep valleys.
- **Mudflows**—these are highly fluid, high-velocity mixtures of sediment and water that have a consistency of wet concrete. Mass wasting of this type typically occurs as certain arid or semiarid mountainous regions are subjected to unusually heavy rains.
- **Earthflows**—these are usually associated with heavy rains and move at velocities between several centimeters and hundreds of meters per year. They usually remain active for long periods of time. They generally tend to be narrow tonguelike features that begin at a scarp or small cliff.

**Slow Movements**

- **Soil creep**—this continuous movement, usually so slow as to be imperceptible, normally occurs on almost all slopes that are moist but not steep enough for landslides. Soil creep is usually accelerated by frost wedging, alternate thawing and freezing, and by certain plant and animal activities. Evidence for creep is often seen in bent trees, offsets in roads and fences, and inclined utility poles.
- **Solifluction**—this downslope movement is typical of areas where the ground is normally frozen to considerable depth—arctic, subarctic, and high mountain regions. The actual soil flowage occurs when the upper portion of the mantle rock thaws and becomes water saturated. The underlying, still frozen subsoil acts as a slide for the sodden mantle rock, which will move down even the gentlest slope.

## Did You Know?

Landslides constitute a major geologic hazard because they are widespread, occur in all fifty states and U.S. territories, and cause $1–2 billion in damages and more than twenty-five fatalities on average each year. Expansion of urban and recreational developments into hillside areas leads to more people that are threatened by landslides each year. Landslides commonly occur in connection with other major natural disasters such as earthquakes, volcanoes, wildfires, and floods (USGS, 2008).

# Desertification

Deserts are areas where the amount of precipitation received is less than the potential evaporation (less than ten inches per year); they cover roughly 30 percent of the earth's land surface—areas we think of as arid. *Desertification* occurs in hot areas far from sources of moisture, areas isolated from moisture by high mountains, in coastal areas along which there are onshore winds and cold-water currents, and/or high-pressure areas where descending air masses produce warm, dry air.

According to USGS (1997), the world's great deserts were formed by natural processes interacting over long intervals of time. During most of these times, deserts have grown and shrunk independent of human activities. Desertification does not occur in linear, easily mappable patterns. Deserts advance erratically, forming patches on their borders. Scientists question whether desertification, as a process of global change, is permanent or how and when it can be halted or reversed.

# Chapter Review Questions

9.1   Wind is more of a _____ of sediment than a creator of sediment.
9.2   Sediment near the ground surface is transported by wind in a process called _____.
9.3   The process of _____ is the lowering of the land surface due to removal of fine-grained particles by the wind.
9.4   A _____ is any bedrock surface or stone or pebble that has been abraded or shaped by wind-blown sediment in a process similar to sandblasting.
9.5   A dune is a major type of wind-blown _____ deposit.
9.6   Crescent-shaped dunes: _____.
9.7   Parabolic dunes are _____-shaped.
9.8   Yellowish, fine-grained, nonstratified wind-carried material: _____.
9.9   _____ takes place when earth materials move downslope in response to gravity.
9.10  _____ are special landslides that occur along a curved surface.

# References and Recommended Reading

Barnes, F. A. 1987. *Canyon Country Arches and Bridges*. Self-published.
Goodwin, P. H. 1998. *Landslides, Slumps, and Creep*. New York: Franklin Watts.
Jennings, T. 1999. Landslides and Avalanches. North Mankato, MN: Thame-side Press.
USGS. 1997. Desertification. Accessed 7/08/08 at http://pubs.usgs.gov/gip/deserts/desertification/.
USGS. 2008. *Landslide Hazards Program*. Accessed 7/22/08 at http://landslides.usgs.gov/.
Vreeland, R. H. 1994. *Nature's Bridges and Arches*, Volume 1, *General Information*, 2nd ed. Self-published.
Walker, J. 1992. *Avalanches and Landslides*. New York: Gloucester Press.

# CHAPTER 10

# Oceans and Their Margins

## Oceans

Oceans are the storehouse of Earth's water. Oceans cover about 71 percent of Earth's surface. Average depth of Earth's oceans is about 3,800 meters, with the greatest ocean depth recorded at 11,036 meters in the Mariana Trench. At the present time, the oceans contain a volume of about 1.35 billion cubic kilometers (96.5 percent of Earth's total water supply), but the volume fluctuates with the growth and melting of glacial ice.

Composition of ocean water has remained constant in composition throughout geologic time. The major constituents dissolving in ocean water (from rivers and precipitation and the result of weathering and degassing of the mantle by volcanic activity) is composed of about 3.5 percent, by weight, of dissolved salts, including chloride (55.07 percent), sodium (30.62 percent), sulfate (7.72 percent), magnesium (3.68 percent), calcium (1.17 percent), potassium (1.10 percent), bicarbonate (0.40 percent), bromine, (0.19 percent) and strontium (0.02 percent).

The most significant factor related to ocean water that everyone is familiar with is the salinity of the water—how salty it is. *Salinity*, a measure of amount of dissolved ions in the oceans, ranges between 33 and 37 parts per thousand. Often the concentration is the amount (by weight) of salt in water, as expressed in "parts per million" (ppm). Water is saline if it has a concentration of more than 1,000 ppm of dissolved salts; ocean water contains about 35,000 ppm of salt (USGS, 2007). Chemical precipitation, absorption onto clay minerals and plants and animals, prevents seawater from containing even higher salinity concentrations. However, salinity does vary in the oceans because surface water evaporates, rain and stream water is added, and ice forms or thaws.

## Did You Know?

Salinity is higher in mid-latitude oceans because evaporation exceeds precipitation. Salinity is also higher in restricted areas of the oceans such as the Red Sea (up to forty parts per thousand). Salinity is lower near the equator because precipitation is higher and is also lower near the mouths of major rivers because of input of fresh water.

Along with salinity, another important property of seawater includes temperature. The temperature of surface seawater varies with latitude, from near 0C near the poles to 29°C near the equator. Some isolated areas can have temperatures up to 37°C. Temperature decreases with ocean depth.

## THE OCEAN FLOOR

The bottoms of the ocean basins (ocean floors) are marked by mountain ranges, plateaus, and other relief features similar to (although not as rugged as) those on the land.

As shown in figure 10.1, the floor of the ocean has been divided into four pieces: the continental shelf, continental slope, continental rise, and deep-sea floor or abyssal plain.

- **Continental shelf**—this is the flooded, nearly flat true margins of the continents. Varying in width to about forty miles and a depth of approximately 650 feet, continental shelves slope gently outward from the shores of the continents (see figure 10.1). Continental shelves occupy approximately 7.5 percent of the ocean floor.
- **Continental slope**—this is a relatively steep slope descending from the continental shelf (see figure 10.1) rather abruptly to the deeper parts of the ocean. These slopes occupy about 8.5 percent of the ocean floor.
- **Continental rise**—this is a broad gentle slope below the continental slope containing sediment that has accumulated along parts of the continental slope.
- **Abyssal plain**—this is a sediment-covered deep-sea plain about 12,000–18,000 feet below sea level. This plane makes up about 42 percent of the ocean floor.

The deep ocean floor does not consist exclusively of the abyssal plain. In places there are areas of considerable relief. Among the more important such features are:

- **Seamounts**—these are isolated mountain-shaped elevations more than 3,000 feet high.
- **Mid-oceanic ridge**—these are submarine mountains, extending more than 37,000 miles through the oceans, generally 10,000 feet above the abyssal plain.

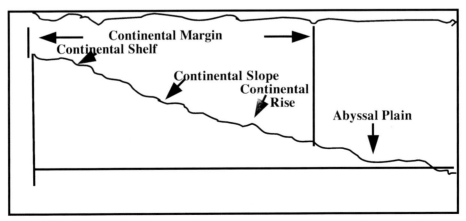

**Figure 10.1.** Cross section of ocean floor showing major elements of topography.

- **Trench**—this is a deep, steep-sided trough in an abyssal plain.
- **Guyot**—this is a seamount that is flat-topped and was once a volcano. They rise from the ocean bottom and usually are covered by 3,000 to 6,000 feet of water.

## OCEAN TIDES, CURRENTS, AND WAVES

Water is the master sculptor of Earth's surfaces. The ceaseless, restless motion of the sea is an extremely effective geologic agent. Besides shaping inland surfaces, water sculpts the coast. Coasts include sea cliffs, shores, and beaches. Seawater set in motion erodes cliffs, transports eroded debris along shores, and dumps it on beaches. Therefore most coasts retreat or advance. In addition to the unceasing causes of motion—wind, density of seawater, and rotation of the earth—the chief agents in this process are tides, currents, and waves.

### Tides

The periodic rise and fall of the sea (once every twelve hours and twenty-six minutes) produces the tides. Tides are due to the gravitational attraction of the moon and to a lesser extent, the sun on the Earth. The moon has a larger effect on tides and causes the Earth to bulge toward the moon. It is interesting to note that at the same time the moon causes a bulge on Earth, a bulge occurs on the opposite side of the Earth due to inertial forces (further explanation is beyond the scope of this text). The effect of the tides is not too noticeable in the open sea, the difference between high and low tide amounting to about two feet. The tidal range may be considerably greater near shore, however. It may range from less than two feet to as much as fifty feet. The tidal range will vary according to the phase of the moon and the distance of the moon from the Earth. The type of shoreline and the physical configuration of the ocean floor will also affect the tidal range.

### Currents

The oceans have localized movements of masses of seawater called ocean currents. These are the result of drift of the upper 50 to 100 meters of the ocean due to drag by wind. Thus, surface ocean currents generally follow the same patterns as atmospheric circulation with the exception that atmospheric currents continue over the land surface while ocean currents are deflected by the land. Along with wind action, current may also be caused by tides, variation in salinity of the water, rotation of the earth, and concentrations of turbid or muddy water. Temperature changes in water affect water density, which, in turn, causes currents—these currents cause seawater to circulate vertically.

### Waves

Waves, varying greatly in size, are produced by the friction of wind on open water. Wave height and power depend upon wind strength and fetch—the amount of unobstructed ocean over which the wind has blown. In a wave, water travels in loops. Essentially an up-and-down movement of the water, the diameter of the loops decreases with depth. The diameter of loops at the surface is equal to wave height (h). Breakers are formed when the wave comes into shallow water near the shore. The lower part of the wave is retarded by the ocean bottom, and the top, having greater momentum, is hurled forward, causing the wave to break. These breaking waves may do great damage to coastal property as they race across coastal lowlands driven by winds of gale or hurricane velocities.

# Coastal Erosion, Transportation, and Deposition

The geologic work of the sea, like previously discussed geologic agents, consists of erosion, transportation, and deposition. The sea accomplishes its work of coastal landform sculpting largely by means of waves and wave-produced currents; their effect on the seacoast may be quite pronounced. The coast and accompanying coastal deposits and landform development represent a balance between wave energy and sediment supply.

## WAVE EROSION

Waves attack shorelines and erode by a combination of several processes. The resistance of the rocks composing the shoreline and the intensity of wave action to which it is subjected are the factors that determine how rapidly the shore will be eroded. Wave erosion works chiefly by hydraulic action, corrosion, and attrition. As waves strike a sea cliff, *hydraulic action* crams air into rock crevices putting tremendous pressure on the surrounding rock; as waves retreat, the explosively expanding air enlarges cracks and breaks off chunks of rock (*scree*). Chunks hurled by waves against the cliff break off more scree (via a sandpapering action)—a process called *corrasion*. The sea rubs and grinds rocks together forming scree that is thrown into the cliffs reducing broken rocks to pebbles and sand grains—a process called *attrition* (Lambert 2007).

Several features are formed by marine erosion—different combinations of wave action, rock type, and rock beds produce these features. Some of the more typical erosion-formed features of shorelines are discussed below.

- **Sea cliffs or wave-cut cliffs**—these are formed by wave erosion of underlying rock followed by the caving-in of the overhanging rocks. As waves eat farther back inland, they leave a wave-cut beach or platform. Such cliffs are essentially vertical and are common at certain localities along the New England and Pacific coasts of North American.
- **Wave-cut bench**—these are the result of wave action not having enough time to

lower the coastline to sea level. Because of the resistance to erosion, a relatively flat wave-cut bench develops. If subsequent uplift of the wave-cut bench occurs, it may be preserved above sea level as a wave-cut bench.

• **Headlands**—these are fingerlike projections of resistant rock extending out into the water. Indentations between headlands are termed *coves*.
• **Sea caves, sea arches, and stacks**—these are formed by continued wave action on a sea cliff. Wave action hollows out cavities or caves in the sea cliffs. Eventually, waves may cut completely through a headland to form a sea arch; if the roof of the arch collapses, the rock left separated from the headland is called a stack.

## MARINE TRANSPORTATION

Waves and currents are important transporting agents. Rip currents and undertow carry rock particles back to the sea, and long-shore currents will pick up sediments (some of it in solution), moving them out from shore into deeper water. Materials carried in solution or suspension may drift seaward for great distances and eventually be deposited far from shore. During the transportation process, sediments undergo additional erosion, becoming reduced in size.

## MARINE DEPOSITION

Marine deposition takes place whenever currents and waves suffer reduced velocity. Some rocks are thrown up on the shore by wave action. Most of the sediments thus deposited consist of rock fragments derived from the mechanical weathering of the continents, and they differ considerably from terrestrial or continental deposits. Due to input of sediments from rivers, deltas may form; due to beach drift, such features as spits and hooks, bay barriers, and tombolos may form. Depositional features along coasts are discussed below.

• **Beaches**—these are transitory coastal deposits of debris that lie above the low-tide limit in the shore zone.
• **Barrier islands**—these are long, narrow accumulations of sand lying parallel to the shore and separated from the shore by a shallow lagoon.
• **Spits and hooks**—these are elongated, narrow embankments of sand and pebbles extending out into the water but attached by one end to the land.
• **Tombolos**—these are bars of sand or gravel connecting an island with the mainland or another island.
• **Wave-built terraces**—these are structures built up from sediments deposited in deep water beyond a wave-cut terrace.
• **Deltas**—these form where sediment supply is greater than ability of waves to remove sediment.

# Chapter Review Questions

10.1 _____ is a measure of the amount of dissolved ions in the oceans.

10.2 _____ is a deep, steep-sided trough in an abyssal plain.

10.3 _____ is a type of seamount that is flat-topped and once was a volcano.

10.4 _____ are due to the gravitational attraction of the moon.

10.5 _____ are fingerlike projections of resistant rock extending out into the water.

# References and Recommended Reading

Gross, G. M. 1995. *Oceanography: A View of the Earth*. Englewood Cliffs, NJ: Prentice-Hall, Inc.

Lambert D. 2007. *The Field Guide to Geology*. New York: Checkmark Books.

Pinet, P. R. 1996. *Invitation to Oceanography*. St. Paul, MN: West Publishing Company.

USGS. 2007. *The Water Cycle: Water Storage in Oceans*. Accessed 7/11/08 at http://ga.water.usgs.gov/edu/watercycleoceans.html.

# CHAPTER 11

# Lakes

## Lentic Habitat

When we look at a globe of the earth we quickly perceive that ours is a water planet. Water covers most of Earth's surface (about three quarters of it). With all of Earth's water, the irony is that if all of Earth's 325 trillion gallons were squeezed into a gallon container and we poured off what was not drinkable (polluted, salty, or frozen) we would be left with one drop. Within this one drop, a very small percentage of its whole represents all the freshwater contained in all of the Earth's lakes.

Lakes and ponds are found in many parts of the world and many of them are of great importance to human welfare. The total volume of freshwater contained in Earth's lakes is 91,000 cubic kilometers. More specifically, freshwater contained in lakes represent 0.007 percent of Earth's total water and 0.26 percent of Earth's total freshwater supply (Spellman, 2008).

Before briefly discussing the geology of lakes and ponds it is important to be familiar with a few key terms.

**Lake**—may be defined as a body of standing surface-water runoff (and maybe some groundwater seepage) occupying a depression in the land.

**Limnology**—is the study of freshwater ecology, which is divided into two classes: lentic and lotic.

- **Lentic class** (calm zone)—is lakes, ponds, and swamps. These are composed of four zones: littoral, limnetic, profundal, and benthic zones.

  - Littoral—the outermost shallow region of the lentic class, which has light penetration to the bottom.
  - Limnetic—the open water zone of the lentic class to a depth of effective light penetration.
  - Euphotic—refers to all lighted regions (light penetration) formed of the littoral and limnetic zones.
  - Profundal—a deep-water region beyond light penetration of the lentic class.
  - Benthic—the bottom region of a lake.

- **Lotic class** (washed)—consists of rivers and streams and is composed of two zones: rapids and pools.

  - In the rapids zone, the stream velocity prevents sedimentation, with a firm

bottom provided for organisms specifically adapted to live attached to the substrate.
- The pool area is a deeper region with a slow enough velocity to allow sedimentation. The bottom is soft due to silts and settleable solids that cause lowered dissolved oxygen due to decomposition.

- **Pond**—is smaller than a lake. Ponds may be natural or man-made in origin.
- **Stratified lake**—can be divided into three horizontal layers: epilimnion (upper, usually oxygenated layer); mesolimnion, or hermocline (middle layer of rapidly changing temperature); and hypolimnion (lowest layer, which is subject to deoxygenation).

## Did You Know?

It's not that water that forms lakes gets trapped, but that water entering a lake comes in faster than it can escape, either via outflow in a river, seepage into the ground, or by evaporation (USGS, 2008).

Lakes and ponds range in size of just a few square feet (ponds are generally two to eight hectares) to thousands of square miles. Scattered throughout the earth, many of the first lakes evolved during the Pleistocene Ice Age. Lakes are found at all altitudes. Lake Titicaca (Peru and Chile) is 12,500 feet above sea level. At the other extreme, the Dead Sea in Israel and Jordan is almost 1,300 feet below sea level. Many ponds are seasonal, such as sessile pools, just lasting a couple of months, while lakes last many years (none—even the Great Lakes—will last forever).

Lakes and ponds are divided into four different "zones," which are usually determined by depth and distance from the shoreline. The four distinct zones—littoral, limnetic, profundal, and benthic—are shown in figure 11.1. Miller (1998) points out that each zone provides a variety of ecological niches for different species of plant and animal life.

The littoral zone is the top-most zone near the shores of the lake or pond, with light penetration to the bottom. It provides an interface zone between the land and the open water of lakes. This zone contains rooted vegetation such as grasses, sedges, rushes, water lilies and waterweeds, and a large variety of organisms. The littoral zone is further divided into concentric zones, with one group replacing another as the depth of water changes. Figure 11.1 also shows these concentric zones: emergent vegetation, floating leaf vegetation, and submerged vegetation zones, proceeding from shallow to deeper water.

The littoral zone is the warmest zone, since it is the area that light hits; it contains flora such as rooted and floating aquatic plants and a very diverse community, which can include several species of algae (like diatoms), grazing snails, clams, insects, crustaceans, fishes, and amphibians. The aquatic plants aid in providing support by estab-

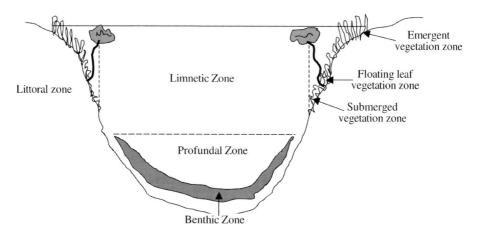

**Figure 11.1.    Vertical section of a pond showing major zones of life (Source: Modified from Enger, Kormelink, Smith, and Smith, 1989, *Environmental Science: An Introduction*. Dubuque, Iowa: Wm. C. Brown Publications, p. 77).**

lishing excellent habitats for photosynthetic and heterotrophic (requiring organic food from the environment) microflora as well as many zooplankton and larger invertebrates (Wetzel 1983).

From figure 11.2 it can be seen that the limnetic zone is the open-water zone up to the depth of effective light penetration; that is, the open water away form the shore. The community in this zone is dominated by minute suspended organisms, the plankton, such as phytoplankton (plants) and zooplankton (animals), and some consumers such as insects and fish. Plankton are small organisms that can feed and reproduce on their own and serve as food for small chains.

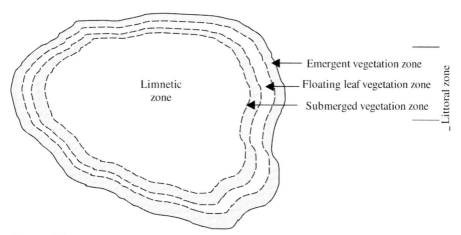

**Figure 11.2.    View looking down on concentric zones that make up the littoral zone.**

## Did You Know?

Without plankton in the water, there would not be any living organisms in the world, including humans.

In the limnetic zone, the population density of each species is quite low. The rate of photosynthesis is equal to the rate of respiration; thus, the limnetic zone is at compensation level. Small shallow ponds do not have this zone; they have a littoral zone only. When all lighted regions of the littoral and limnetic zones are discussed as one, the term euphotic is used for both, designating these zones as having sufficient light for photosynthesis and the growth of green plants to occur.

The small plankton do not live for a long time. When they die, they fall into the deepwater part of the lake/pond, the profundal zone. The profundal zone, because it is the bottom, or deepwater region, is not penetrated by light. This zone is primarily inhabited by heterotrophs adapted to its cooler, darker water and lower levels of oxygen.

The final zone, the benthic zone, is the bottom region of the lake. It supports scavengers and decomposers that live on sludge. The decomposers are mostly large numbers of bacteria, fungi, and worms, which live on dead animal and plant debris and wastes that find their way to the bottom.

## Classification of Lakes

It is our natural tendency to classify things. In regards to lakes, Kevern et al. (1999) use three classifications. One is based on productivity of the lake (or its relative richness). This is the trophic basis of classification. A second classification is based on the times during the year that the water of a lake becomes mixed and the extent to which the water is mixed. And a third classification is based on the fish community of lakes.

For the purpose of this text, a somewhat different classification scheme is used rather than the one just described. That is, lakes are classified based on eutrophication, special types of lakes, and impoundments.

**Eutrophication** is a natural aging process that results in organic material being produced in abundance due to a ready supply of nutrients accumulated over time. Through natural succession (i.e., the process by which biological communities replace each other in a relatively predictable sequence), eutrophication causes a lake ecosystem to turn into a bog and eventually to a terrestrial ecosystem. Eutrophication has received a great amount of publicity lately. In recent years, humans have accelerated the eutrophication of many surface waters by the addition of wastes containing nutrients. This accelerated process is called cultural eutrophication. Sources of human wastes and pollution are sewage, agricultural runoff, mining, industrial wastes, urban runoff, leaching from cleared land, and landfills.

## CLASSIFICATION OF LAKES BASED ON EUTROPHICATION

Lakes can be classified into three types based on their eutrophication stage.

1. **Oligotrophic lakes** (few foods): These are young, deep, crystal-clear, nutrient-poor lakes with little biomass productivity. Only a small quantity of organic matter grows in an oligotrophic lake; the phytoplankton, the zooplankton, the attached algae, the macrophytes (aquatic weeds), the bacteria, and the fish are all present as small populations. It's like planting corn in sandy soil, not much growth (Kevern et al., 1999). Lake Superior is an example from the Great Lakes.

2. **Mesotrophic lakes**: It is hard to draw distinct lines between oligotrophic and eutrophic lakes, and often the term mesotrophic is used to describe a lake that falls somewhere between the two extremes. Mesotrophic lakes develop with the passage of time. Nutrients and sediments are added through runoffs, and the lake becomes more productive biologically. There is a great diversity of species with very low populations at first, but a shift toward higher and higher populations of fewer and fewer species. Sediments and solids contributed by runoffs and organisms make the lake shallower. At an advanced mesotrophic stage, a lake has undesirable odors and colors in certain parts. Turbidity increases, and the bottom has organic deposits. Lake Ontario has reached this stage.

(3) **Eutrophic lakes** (good foods): This is a lake with a large or excessive supply of nutrients. As the nutrients continue to be added, large algal blooms occur, fish types change from sensitive to more pollution-tolerant ones, and biomass productivity becomes very high. Populations of a small number of species become very high. The lake takes on undesirable characteristics, such as offensive odors, very high turbidity, and a blackish color. This high level of turbidity can be seen in studies of Lake Washington in Seattle, Washington. Laws (1993) reports that Secchi depth (measure of turbidity of water) measurements made in Lake Washington from 1950 to 1979 show an almost fourfold reduction in water clarity. Along with the increase in turbidity, the lake becomes very shallow. Lake Erie is at this stage. Over a period of time, a lake eventually becomes filled with sediments as it evolves into a swamp and finally into a land area.

## SPECIAL TYPES OF LAKES

Odum (1971) refers to several special lake types.

1. **Dystrophic** (like bog lakes): These lakes develop from the accumulation of organic matter from the surrounding land. In this case, the watershed is often forested and there is an input of organic acids (e.g., humic acids) from the breakdown of leaves and evergreen needles. There follows a rather complex series of events and processes resulting finally in a lake that is usually low in pH (acid) and often is moderately clear, but color ranges from yellow to brown. Dissolved solids, nitrogen, phospho-

rus, and calcium are low, and humic matter is high. These lakes are sometimes void of fish fauna; other organisms are limited. When fish are present, production is usually poor. They are typified by the bog lakes of northern Michigan.

2. **Deep ancient lakes:** These lakes contain animals found nowhere else (endemic fauna). For example, Lake Baikal in Russia.

3. **Desert salt lakes:** These are specialized environments like the Great Salt Lake in Utah, where evaporation rates exceed precipitation rates, resulting in salt accumulation.

4. **Volcanic lakes:** These are lakes on volcanic mountain peaks, as in Japan and the Philippines.

5. **Chemically stratified lakes:** Examples include Big Soda Lake in Nevada. These lakes are stratified due to different densities of water caused by dissolved chemicals. They are meromictic, which means partly mixed.

6. **Polar lakes:** These are lakes in the polar regions, their surface water temperature mostly below 4°C.

7. **Marl lakes:** According to Kevern et al. (1999), Marl lakes are different in that they generally are very unproductive, yet they may have summertime depletion of dissolved oxygen in the bottom waters and very shallow Secchi disk depths, particularly in the late spring and early summer. These lakes gain significant amounts of water from springs that enter at the bottom of the lake. When rainwater percolates through the surface soils of the drainage basin, the leaves, grass, and other organic materials incorporated in these soils are attacked by bacteria. These bacteria extract the oxygen dissolved in the percolating rainwater and add carbon dioxide. The resulting concentrations of carbon dioxide can get quite high, and when they interact with the water, carbonic acid is formed.

   As this acid-rich water percolates through the soils, it dissolves limestone. When such groundwater enters a lake through a spring, it contains very low concentrations of dissolved oxygen and is supersaturated with carbon dioxide. The limestone that was dissolved in the water reforms very small particles of solid limestone in the lake as the excess carbon dioxide is given off from the lake to the atmosphere. These small particles of limestone are marl, and, when formed in abundance, cause the water to appear turbid, yielding a shallow Secchi disk depth. The low dissolved oxygen in the water entering from the springs produces low dissolved oxygen concentrations at the lake bottom.

## IMPOUNDMENTS (SHUT-INS)

These are artificial lakes made by trapping water from rivers and watersheds. They vary in their characteristics according to the region and nature of drainage. They have high turbidity and a fluctuating water level. The biomass productivity, particularly of benthos, is generally lower than that of natural lakes (Odum 1971).

# Chapter Review Questions

11.1  A _____ is a body of standing surface-water runoff occupying a depression in the land.

11.2  Lentic class includes _____, _____, and _____.

11.3  The _____ zone is a deep-water region beyond light penetration in a lake.

11.4  The _____ zone consists of the rapids and pool zones.

11.5  A _____ lake is characterized as possessing a large or excessive supply of nutrients.

# References and Recommended Reading

Kevern, N. R., King, D. L., and Ring, R. 1999. Lake classification systems. Part I. *The Michigan Riparian* (December), p. 1.

Laws, E. A. 1993. *Aquatic Pollution: An Introductory Text.* New York: Wiley, p. 59.

Miller, G. T. 1998. *Environmental Science: An Introduction.* Belmont, CA: Wadsworth Publishing, p. 77.

Odum, E. P. 1971. *Fundamentals of Ecology.* Philadelphia: Saunders, pp. 312–13.

Spellman, F. R. 2008. *The Science of Water*, 2nd ed. Boca Raton, FL: CRC Press.

USGS. 2008. *Earth's Water: Lakes and Reservoirs.* Accessed 09/14/08 at http://ga.water.usgs.gov/edu/earthlakes.html.

Wetzel, R. G. 1983. *Limnology.* New York: Harcourt College.

## CHAPTER 12

# Earthquakes

*It's been raining a lot, or very hot—it must be earthquake weather!*

Fiction: Many people believe that earthquakes are more common in certain kinds of weather. In fact, no correlation with weather has been found. Earthquakes begin many kilometers (miles) below the region affected by surface weather. People tend to notice earthquakes that fit the pattern and forget the ones that don't. Also, every region of the world has a story about earthquake weather, but the type of weather is whatever they had for their most memorable earthquake.

—USGS (2008)

## What Causes Earthquakes

Anyone who has witnessed (been exposed to) or studied one of the million or so earthquakes that occur each year on Earth is unlikely to forget such occurrences. Even though most earthquakes are insignificant, a few thousand of these produce noticeable effects, such as tremors and/or ground shaking. The passage of time has shown that about twenty earthquakes each year cause major damage and destruction. It is estimated that about 10,000 people die each year because of earthquakes.

Over the millennia, the effects of damaging earthquakes have been obvious to those who witnessed the results. However, the cause of earthquakes has not been as obvious. For example, the cause of earthquakes has shifted from the incantations of mythical beasts, to the wrath of gods, to unexplainable magical occurrences, to just normal, natural phenomena occasionally required to retain Earth's structural integrity, providing Earth with a periodic form of feedback to keep "things" in balance. We can say, overall, that an earthquake provides our planet with a sort of a geological homeostasis needed to maintain life as we know it.

Through the ages, earthquakes have also come under the attention and eventually the pen of some of the world's greatest writers. Consider, for example, Voltaire's classic satirical novel, *Candide*, published in 1759, in which he mercilessly satirizes the Enlightenment attitude of optimism by weighing in on the 1755 great earthquake in Lisbon, Portugal, which is believed to have caused the deaths of more than 60,000 people. Dr. Pangloss says to Candide (on viewing the total devastation of Lisbon):

"The heirs of the dead will benefit financially; the building trade will enjoy a boom. Private misfortune must not be overrated. These poor people in

163

their death agonies, and the worms about to devour them, are playing their proper and appointed part in God's master plan."

Although we still do not know what we do not know about earthquakes and their causes, we have evolved from using witchcraft or magic to explain their origins to the scientific methods employed today. In the first place, we do know that earthquakes are caused by the sudden release of energy along faults. Earthquakes are usually followed by a series of smaller earthquakes that we called **aftershocks**. Aftershocks represent further adjustments of rock along the fault. There are currently no reliable methods for predicting when earthquakes will occur.

In regards to the cause(s) or origin(s) of earthquakes, we have developed a couple of theories. One of these theories explains how earthquakes occur via *elastic rebound*. That is, according to elastic rebound theory, subsurface rock masses subjected to prolonged pressures from different directions will slowly bend and change shape. Continued pressure sets up strains so great that the rocks will eventually reach their elastic limit and rupture (break) and suddenly snap back into their original unstrained state. It is the snapping back (elastic rebound) that generates the seismic waves radiating outward from the break. The greater the stored energy (strain), the greater the release of energy.

The coincidence of many active volcanic belts with major belts of earthquake activity (*seismic* and *volcanic activity*) indicates that volcanoes and earthquakes may have a common cause. Plate interactions commonly cause both earthquakes (tectonic earthquakes) and volcanoes.

# Seismology

Even though *seismology* is the study of earthquakes, it is actually the study of how seismic waves behave in the earth. The source of an earthquake is called the hypocenter or **focus** (i.e., the exact location within the earth where seismic waves are generated). The **epicenter** is the point on the earth's surface directly above the focus. Seismologists want to know where the focus and epicenter are located so a comparative study of the behavior of the earthquake event can be made with previous events—in an effort to further understanding.

Seismologists use instruments to detect, measure, and record seismic waves. Generally, the instrument used is the **seismograph**, which has been around for a long time. Modern updates have upgraded these instruments from the paper or magnetic tape strip to electronically recorded data that is input to a computer. A study of the relative arrival times of the various types of waves at a single location can be used to determine the distance to the epicenter. To determine the exact epicenter location, records from at least three widely separated seismograph stations are required.

# Seismic Waves

As mentioned, some of the energy released by an earthquake travels through the earth. The speed of a seismic wave depends on the density and elasticity of the materials through which it travels. Seismic waves come in several types as described below:

- **P-waves**—primary, pressure, or push-pull waves (the first detected by seismograph) are compressional waves (expanding and contracting) that travel through the earth (solids, liquids, or gases) at speeds of from 3.4 to 8.6 miles per second. P-waves move faster at depth, depending on the elastic properties of the rock through which they travel. P-waves are the same thing as sound waves.
- **S-waves**—secondary or shear waves travel with a velocity (between 2.2 and 4.5 miles per second) that depends only on the rigidity and density of the material through which they travel. They are the second set of waves to arrive at the seismograph and will not travel through gases or liquids; thus the velocity of S-waves through gas or liquids is zero.
- **Surface waves**—Several types travel along the earth's outer layer or surface or on layer boundaries in the earth. These are rolling, shaking, waves that are the slowest waves but the ones that do the damage in large earthquakes.

# Earthquake Magnitude and Intensity

The size of an earthquake is measured using two parameters—energy released (magnitude) and damage caused (intensity).

## EARTHQUAKE MAGNITUDE

The size of an earthquake is usually given in terms of its Richter magnitude. Richter magnitude is a scale devised by Charles Richter that measures the amplitude (height) of the largest recorded wave at a specific distance from the earthquake. A better measure is the Richter scale, which measures the total amount of energy released by an earthquake as recorded by seismographs. The amount of energy released is related to the Richter scale by the equation:

$$\text{Log E} = 11.8 + 1.5 \text{ M}$$

where
    Log = the logarithm to the base 10
    E = the energy released in ergs
    M = is the Richter magnitude

In using the equation to calculate Richter magnitude, it quickly becomes apparent that we see that each increase of 1 in Richter Magnitude yields a 31-fold increase in the amount of energy released. Thus, a magnitude 6 earthquake releases 31 times more energy than a magnitude 5 earthquake. A magnitude 9 earthquake releases 31 × 31 or 961 times more energy than a magnitude 7 earthquake.

# Did You Know?

While it is correct to say that for each increase of 1 in the Richter magnitude, there is a tenfold increase in amplitude of the wave, it is incorrect to say that each increase of 1 in Richter magnitude represents a tenfold increase in the size of the earthquake.

## EARTHQUAKE INTENSITY

Earthquake intensity is a rough measure of an earthquake's destructive power (i.e., size and strength—how much the earth shook at a given place near the source of an earthquake). To measure earthquake intensity, Mercalli in 1902 devised an intensity scale of earthquakes based on the impressions of people involved, movement of furniture and other objects, and damage to buildings. The shock is most intense at the epicenter, which, as noted earlier, is located on the surface directly above the focus.

Mercalli's intensity scale uses a series of numbers (based on a scale of 1 to 12) to indicate different degrees of intensity (see table 12.1). Keep in mind that this scale is somewhat subjective, but it provides a qualitative, but systematic, evaluation of earthquake damage.

# Internal Structure of the Earth

Information obtained from seismographs and other instruments indicate that the lithosphere may be divided into three zones: the crust, mantle, and core (see figure 12.1).

**Table 12.1.    Modified Mercalli Intensity Scale**

| Intensity | Description |
| --- | --- |
| I | Not felt except under unusual conditions |
| II | Felt by only a few on upper floors |
| III | Felt by people lying down or seated |
| IV | Felt indoors by many, by few outside |
| V | Felt by everyone, people awakened |
| VI | Trees sway, bells ring, some objects fall |
| VII | Causes alarm, walls and plaster crack |
| VIII | Chimneys collapse, poorly constructed buildings seriously damaged |
| IX | Some houses collapse, pipes break |
| X | Ground cracks, most buildings collapse |
| XI | Few buildings survive, bridges collapse |
| XII | Total destruction |

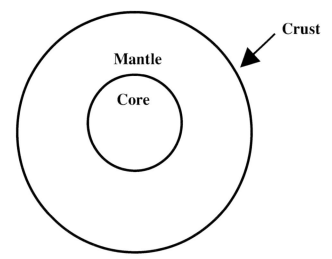

**Figure 12.1.  Internal structure of Earth.**

- **Earth's crust:** The outermost and thinnest layer of the lithosphere is called the crust. There are two different types of crust: thin (as little as four miles in places) oceanic crust (compose primarily of basalt) that underlies the ocean basins, and thicker continental crust (primarily granite twenty to thirty miles thick) that underlies the continents.
- **Earth's mantle:** Beneath the crust is an 1,800-mile-thick intermediate, dense, hot zone of semisolid rock called the mantle. It is thought to be composed mainly of olivine-rich rock.
- **Earth's core:** Earth's core is about 4,300 miles in diameter. It is thought to be composed of a very hot, dense iron and nickel alloy. The core is divided into two different zones. The outer core is a liquid because the temperatures there are adequate to melt the iron-nickel alloy. The highly pressurized inner core is solid because the atoms are tightly crowded together.

# Chapter Review Questions

12.1 According to the _____ _____ theory, subsurface rock masses subjected to prolonged pressures from different direction will slowly bend and change shape.

12.2 Plate _____ commonly cause both earthquakes and volcanoes.

12.3 _____ is the study of earthquakes.

12.4 To determine the exact epicenter location of an earthquake, records from at least _____ widely separated seismograph stations are required.

12.5 _____ is the measure of damage caused by an earthquake.

# References and Recommended Reading

USGS. 2008. *Earthquake Hazards Program*. Accessed 09/14/08 at http://earthquakes.usgs.gov/ learning/topics/megaqk_facts_fantasy.php.

Voltaire. 1991. *Candide*. New York: Dover.

## CHAPTER 13

# Plate Tectonics

Within the past forty-five or fifty years geologists have developed the theory of plate tectonics (tectonics: Greek, "builder"). The theory of plate tectonics deals with the formation, destruction, and large scale motions of great segments of Earth's surface (crust), called **plates**. This theory relies heavily on the older concepts of continental drift (developed during the first half of the twentieth century), and seafloor spreading (understood during the 1960s), which help to explain the cause of earthquakes and volcanic eruptions, and the origin of fold mountain systems.

## Crustal Plates

Earth's crustal plates are composed of great slabs of rock (lithosphere), about 100 kilometers thick and covering many thousands of square miles (they are thin in comparison to their length and width); they float on the ductile **asthenosphere**, carrying both continents and oceans. Many geologists recognize at least eight main plates and numerous smaller ones. These main plates include

- African plate, covering Africa—continental plate
- Antarctic plate, covering Australia—continental plate
- Australian plate, covering Australia—continental plate
- Eurasian plate, covering Asia and Europe—continental plate
- Indian plate, covering Indian subcontinent and a part of Indian Ocean—continental plate
- Pacific plate, covering the Pacific Ocean—oceanic plate
- North American plate, covering North America and northeast Siberia—continental plate
- South American plate, covering South America—continental plate

The minor plates include

- Arabian plate
- Caribbean plate
- Juan de Fuca plate
- Cocos plate
- Nazea plate
- Philippine plate
- Scotia plate

# Plate Boundaries

As mentioned, the asthenosphere is the ductile, soft, plastic-like zone in the upper mantle on which the crustal plates ride. Crustal plates move in relation to one another at one of three types of plate boundaries: **convergent** (collision boundaries), **divergent** (spreading boundaries), and **transform** boundaries. These boundaries between plates are typically associated with deep-sea trenches, large faults, fold mountain ranges, and mid-ocean ridges.

## CONVERGENT BOUNDARIES

Convergent boundaries (or active margins) develop where two plates slide toward each other, commonly forming either a subduction zone (if one plate subducts or moves underneath the other) or a continental collision (if the two plates contain continental crust). To relieve the stress created by the colliding plates, one plate is deformed and slips below the other.

## DIVERGENT BOUNDARIES

Divergent boundaries occur where two plates slide apart from each other. Mid-ocean ridges, which are examples of these divergent boundaries, are where new oceanic, melted lithosphere materials well up, resulting in basaltic magmas that intrude and erupt at the ridge, in turn creating new oceanic lithosphere and crust (new ocean floor). Along with volcanic activity, the mid-ocean ridges are also areas of seismic activity.

## TRANSFORM PLATE BOUNDARIES

Transform, or shear/constructive boundaries, do not separate or collide; rather, they slide past each other in a horizontal manner with a shearing motion. Most transform boundaries occur where mid-ocean ridges are offset on the sea floor. The San Andreas Fault in California is an example of a transform fault.

# Chapter Review Questions

13.1  Earth's _____ are composed of great slabs of rock.
13.2  Plate that covers Pacific Ocean: _____
13.3  _____ boundaries develop where two plates slide toward each other.
13.4  _____ boundaries develop where two plates slide apart from each other.
13.5  _____ boundaries slide past each other.

# References and Recommended Reading

Holmes, A. 1978. *Principles of Physical Geology*, 3rd, New York: Wiley.

Lyman, J., and Fleming, R. H. 1940. Composition of seawater. *J Mar Res* 3, pp. 134–46.

McKnight, T. 2004. *Geographica: The Complete Illustrated Atlas of the World*. New York: Barnes and Noble Books.

Oreskes, N. (ed). 2003. *Plate Tectonics: An Insiders History of the Modern Theory of the Earth*. New York: Westview.

Stanley, S. M. 1999. *Earth System History*. New York: W. H. Freeman, 211–22.

Sverdrup, H. U., Johnson, M. W., and Fleming, R. H. 1942. *The Oceans: Their Physics, Chemistry and General Biology*. New York: Prentice-Hall.

Turcotte, D. L., and Schubert, G. 2002. *Geodynamics*, 2nd ed. New York: Wiley.

## CHAPTER 14

# Geologic Time

The earth scorns our simplifications, and becomes much more interesting in its derision. The history of life is not a continuum of development, but a record punctuated by brief, sometimes geologically instantaneous, episodes of mass extinction and subsequent diversification. The geologic time scale maps this history, for fossils provide our chief criterion in fixing the temporal order of rocks. . . . Hence, the time scale is not the devil's ploy for torturing students, but the chronicle of key moments in life's history. . . . I make no apologies for the central importance of such knowledge.

—Stephen J. Gould

All things continue as they were from the beginning of the creation.

—II Peter 3:3–6

In the study of geology, it is important to give some thought to historical geology, how geologists deal with time, and to the age of Earth. In this text, the focus is on providing basic information on physical and not historical geology. However, it is important to note that historical geology is an important subset field of geology. In historical geology, the principles of geology are used to reconstruct and understand the history of Earth (Levin 2003). Focusing on geologic processes that change the earth's surface and subsurface, historical geology, to explain the sequence of these events, uses stratigraphy, structural geology, and paleontology to explain the sequence of these events. In addition, the evolution of animals and plants during different time periods in the geological timescale is also an area of focus. Radiometric dating techniques provide a means of deriving relative age versus absolute age of geologic history.

- **Relative age (or relative time)**—means that we can determine if something is younger than or older than something else. This is accomplished by dating events in chronologic order of occurrence rather than in years. In other words, relative time gives us the sequence of events but not how old something is. For example, World War I happened before World War II.
- **Absolute age (or absolute time)**—means geologic time measured in a more or less precise time (in years, minutes, seconds, or some other unit of time) to the amount of time that has already passed. In other words, absolute time allows us to say how old something is. For example, the last major ice age occurred 11,000–12,000 years ago.

In geology, we use principles to determine relative ages, correlations, and absolute ages. For relative ages we use stratigraphy. For correlations we use physical criteria, fossils (see figure 14.1) and key beds. For absolute ages we use radiometric dating.

# Stratigraphy

**Stratigraphy** is the geologic subscience dealing with the definition and interpretation of strata (layers) and stratified rocks in the Earth's crust, especially their lithology, sequence, distribution, and correlation.

Earth's layers have been characterized in different ways. One characterization commonly used is to compare Earth to an onion. When we cut an onion across, we find a series of layers surrounding a central core. A cross-section of Earth would show a similar structure. Earth's layers downward are denser and heavier than the layers above.

Another way of characterizing the structure of earth is to compare it to a book. In fact, Saint Augustine characterized Earth as a book (in general and not geological terms) when he said: "The world is a book, and those who do not travel read only one page." We can basically state Augustine's analogy (for the purpose of this text) in a different way—in a "geological" way. We can say Earth's structural layers are like

**Figure 14.1.** *Priscacara serrata*—50-million-year-old Green River fossil fish.

the pages of a book, and when we only study its outer layer, we read only one page. Earth's pages are not new, unread, or in pristine condition, as you would expect in a brand new library book. No, the pages that make up Earth's book are well used, torn, cornered, crumpled, marked, faded in places, turned upside down, displaced, or lost. Earth's bookbinding (along with tremendous levels of pressure) is the force of gravity.

Sedimentary rock layers, or strata, were laid down on top of one another; again, like the pages in a book—Earth's book. Simply, when these pages are correctly read and interpreted, Earth's relative history is revealed. As mentioned, the study (reading) of the layers of Earth is stratigraphy.

## STRATIGRAPHIC LAWS

The reading of Earth's "pages" to determine its relative age is based on the stratigraphic laws. Developed in the seventeenth to nineteenth centuries by Niels Steno, James Hutton, and William Smith, among others, the modern-day geologist uses these principles to decipher the spatial and temporal relationships of rock layers. Basically, geologists studying earth are concerned with the sequence of rocks and structures in time, and thus with the history of the earth itself. It is important to point out, however, that these laws cannot determine the age of earth's layers, simply the relative order in which they were formed. Stratigraphic laws include the following:

- original horizontality
- lateral continuity
- superposition
- cross-cutting relations
- laws of inclusions
- law of faunal succession

## Important Point!

In order to understand the stratigraphic laws, we must assume that the geologic processes of today were the same in the past. We call this *uniformitarianism.*

### Original Horizontality

This law states that all sedimentary rocks are originally deposited horizontally or nearly horizontally. If sedimentary rocks are no longer horizontal, an event occurred subsequent to the deposition that caused the layers to tilt or fold from their original position. In 1669, Nicholas Steno (a Danish geological pioneer) stated this relationship "strata either perpendicular to the horizon or inclined to the horizon were at one time parallel to the horizon" (Levin 2005). The law of original horizontality holds true in the deposition of most sedimentary material; however, one noted exception is sand deposited nonhorizontally at less than fifteen degrees (or at the angle of repose) in sand dunes.

### Lateral Continuity

In this law, Steno (1669) pointed out that sedimentary rocks are laterally continuous over large geographic areas of Earth (deposits originally extended in all directions) unless some other solid bodies stood in the way, or they thin out at edge or grade into a different type of sediment (Levin 2005).

### Superposition

Another stratigraphic law states that because of Earth's gravity, the oldest layer of sediment is at the bottom of the sequence, and the youngest at the top. Thus, in a sequence of layers that has not been disturbed or overturned by a later event, the oldest layers will be on the bottom. Steno stated that "at the time when any given stratum was being formed, all the matter resting upon it was fluid, and, therefore, at the time when the lower stratum was being formed, none of the upper strata existed" (Levin 2005).

### Cross-Cutting Relations

In cross-cutting relations, Steno (1669) points out, "If a body or discontinuity cuts across a stratum, it must have formed after that stratum" (Levin 2005). This law basically describes the relationship between existing rock and rock that intrudes by magma flow into existing rock. This creates an intrusion, and the intrusion is always younger than the rock it invades.

### Law of Inclusions

The law of inclusions states that rock fragments (in other rocks) must be older than the rock containing (surrounding) the fragments.

### Law of Faunal Succession

The law of faunal succession, developed by William Smith (father of English geology, 1769–1839), recognized that fossil faunas (assemblages of animals that lived together at a given time and place) follow (succeeded) each other in a definite and determinable order. These faunas are distinctive for each portion of the earth's history, and allowed geologists to develop a fossil stratigraphy, providing a means to correlate rocks. That is, by comparing these fossils, the geologist is able to recognize deposits of the same ages.

*Structural Breaks in the Stratigraphic Record*

Because of uplift, subsidence, and deformation, the earth's surface is continually changing, with surface erosion acting in some places and erosion of surface sediment occurring in other places. Whenever erosion is removing previous deposited sediment, or when sediment is not being deposited, there will be a break in the continuous record of sedimentation preserved in the rocks. This break in the stratigraphic record is called an unconformity (or hiatus). An **unconformity** is an uplifted surface where there are long periods of erosion or nondeposition. Geologists recognize three basic types of unconformities.

- **Angular unconformity**—this is an easily recognized type of unconformity because the beds above the unconformity are not parallel to the beds below it. This type of unconformity (commonly known as Hutton's (1787) unconformity, indicates that the lower series of rocks were tilted or folded prior to their erosion and the subsequent deposition of the overlying beds.
- **Disconformity**—these are known as parallel unconformities because the rock layers above and below the unconformity are parallel. Because there is often no angular relationship between sets of layers, disconformities are much harder to recognize in the field.
- **Nonconformity**—this type of unconformity is formed when overlying stratified, sedimentary rocks lie on an eroded surface of igneous or metamorphic rocks.

# Relative versus Absolute Time

We cannot study geology without referring to geologic time. Geologic time is often discussed in two forms:

- **Relative time**—a chronostratic arrangement of geologic events and time periods in their proper order (displayed as the Geologic Time Scale; see table 14.1). This is done by using the stratigraphic techniques (relative age relationships—vertical/stratigraphic positions) discussed earlier.
- **Absolute time**—a chronometric arrangement of numerical ages in millions of years or some other measurement; the time in years since the beginning or end of a period. These are commonly obtained via radiometric dating methods performed on appropriate rock types.

## RELATIVE TIME: THE GEOLOGIC COLUMN

The **geologic column** or more commonly known as the standard geologic column is the concept used to break relative geologic time into units of known relative age. In a sense, the column depicts snapshots of geologic time. The geologic column accomplishes this by scaling the oldest to the most recent rocks found in the entire earth or in a given area.

The **geologic time scale** (see table 14.1; based on present state of knowledge) is composed of named intervals of geologic time (relative) during which the rocks represented in the geologic column were deposited.

## Did You Know?

The largest unit of geologic time is an **era**, and each era is divided into smaller time units called **periods**. A period of geologic time is divided into **epochs**, which in turn may be subdivided into still smaller units. The geologic column provides a standard by which we can discuss the relative age of rock formations and the rocks and the fossils they contain. However, again, keep in mind that these time units are arbitrary and of unequal duration, and as we are dealing with relative time, we cannot be positive about the exact amount of time involved in each unit.

### ABSOLUTE TIME: RADIOMETRIC TIME SCALE

Although it is easy to establish the relative ages of rocks based on the work of geologists who have applied the principles of stratigraphy, knowing how much time a geologic era, period, or epoch represents is much more difficult unless we have the absolute ages of rocks. Absolute time measurements are used to calibrate the relative time scale.

In the early years of geology, many attempts were made to establish absolute time measurements. Some of these give only a very rough approximation of the age of rocks; others (radiometric dating) are much more accurate.

- **Salinity of the sea**—the oceans were probably originally composed of fresh water; thus, the age of the earth can be estimated on the basis of how long it would take the oceans to obtain their present salt content.
- **Rate of sedimentation**—if we knew how long it took to deposit all of the rock layers in the crust we could get some idea as to the age of the earth.
- **Radiometric time scale**—this is the most recent and accurate method yet devised for measuring absolute geologic time.

New vistas in science were opened by Henry Becquerel in 1896 when he discovered the natural radioactive decay of uranium. However, it wasn't until 1905 that the British physicist Lord Rutherford suggested using radioactivity as a tool for directly measuring geologic time. Two years later, in 1907, B. B. Boltwood, radiochemist of Yale University, published a list of geologic ages based on radioactivity, which, during the current era, have been modified somewhat to reflect greater accuracy and proper application. Precise dating has been accomplished since 1950 (USGS, 2001).

**Table 14.1.  Geologic Time Scale (USGS 2007; USGS 2008)**

| Erathem or Era | System, Subsystem or Period, Subperiod | Series or Epoch |
|---|---|---|
| **Cenozoic**<br>65 million years ago to present<br>"Age of Recent Life" | **Quaternary**<br>1.8 million years ago to the Present | **Holocene**<br>11,477 years ago (+/- 85 years) to the Present—Greek *holos* ("entire") and *ceno* ("new") |
| | | **Pleistocene**<br>1.8 million to approx. 11,477 (+/- 85 years) years ago—The Great *pleistos* ("most") and *ceno* ("new") |
| | **Tertiary**<br>65.5 to 1.8 million years ago | **Pliocene**<br>5.3 to 1.8 million years ago—Greek *pleion* ("more") and *ceno* ("new") |
| | | **Miocene**<br>23.0 to 5.3 million years ago—Greek *meion* ("less") and *ceno* ("new") |
| | | **Oligocene**<br>33.9 to 23.0 million years ago—Greek *oligos* ("little") and *ceno* ("new") |

**Mesozoic**
251.0 to 65.5 million years ago—Greek means "middle life"

**Cretaceous**
145.5 to 65.5 million years ago

"The Age of Dinosaurs"

Late or Upper

Early or Lower

**Eocene**
55.8 to 33.9 million years ago—Greek *eos* ("dawn") and *ceno* ("new")

**Paleocene**
65.5 to 58.8 million years ago—Greek *palaois* ("old") and *ceno* ("new")

**Jurassic**
199.6 to 145.5 million years ago

Late or Upper

Middle

Early or Lower

**Triassic**
251.0 in 199.6 million years ago

Late or Upper

Middle

Early or Lower

**Table 14.1.** (Continued)

| Era | Period | Epoch / Series |
|---|---|---|
| **Paleozoic** 542.0 to 251.0 million years ago "Age of Ancient Life" | **Permian** 299.0 to 251.0 million years ago | Lopingian |
| | | Guadalupian |
| | | Cisuralian |
| | **Pennsylvanian** 318.1 to 299.0 million years ago "The Coal Age" | Late or Upper |
| | | Middle |
| | | Early or Lower |
| | **Mississippian** 359.2 to 318.1 million years ago | Late or Upper |
| | | Middle |
| | | Early or Lower |
| | **Devonian** 416.0 to 359.2 million years ago | Late or Upper |
| | | Middle |
| | | Early or Lower |

**Silurian**
443.7 to 416.0 million years ago

Pridoli

Early or Lower

Ludlow

Wenlock

Llandovery

**Ordovician**
488.3 to 443.7 million years ago

Late or Upper

Middle

Early or Lower

**Cambrian**
542.0 to 488.3 million years ago

Late or Upper

Middle

Early or Lower

**Precambrian**
approximately 4 billion years ago to 542.0 million years ago

In order to understand the methodology used to construct the radiometric time scale, it is important to know that a chemical element consists of atoms with a specific number of protons in their nuclei, but the number of neutrons varies and therefore elements will have varying atomic weights. Atoms of the same element with differing atomic weights are called isotopes. Isotopes are formed spontaneously when the isotope (the parent) loses particles from its nucleus to form an isotope of a new element (the daughter). The rate of decay is expressed in terms of the time it takes for one-half (an isotope's half-life) of a particular radioactive isotope in a sample to decay. Most radioactive isotopes have short half-lives and lose their radioactivity within a few days or years. However, some isotopes decay slowly, and several of these are used as geologic clocks. The parent isotopes and corresponding daughter products most commonly used to determine the ages of ancient rocks are listed in table 14.2.

USGS (2001) points out that dating rocks by radioactive timekeepers is theoretically simple, but the laboratory procedures are complex. The principal difficulty lies in measuring precisely very small amounts of isotopes.

Because potassium is found in most rock-forming minerals, the potassium-argon method can be used on rocks as young as a few thousand years as well as on the oldest rocks known. The half-life of potassium's radioactive isotope potassium-40 is such that measurable quantities of argon (daughter) have accumulated in potassium-bearing minerals of nearly all ages, and the amounts of potassium and argon isotopes can be measured accurately, even in very small quantities.

## Did You Know?

In practice and where feasible, two or more radioactive dating methods of analysis are used on the same specimen of rock to confirm the results.

Another important isotope used for dating purposes is the carbon-14, which has a half-life of 5,730 years. As a result of the bombardment of nitrogen by neutrons from cosmic rays, the radiocarbon carbon-14 is produced continuously in the earth's upper atmosphere. This radiocarbon becomes uniformly mixed with the nonradioactive carbon in the carbon dioxide of the air, and it eventually finds its way into all

**Table 14.2.   Parent Isotopes and Corresponding Daughter Products (USGS, 2001)**

| Parent Isotope | Stable Daughter Product | Half-Life Values |
| --- | --- | --- |
| Uranium-238 | Lead-206 | 4.5 billion years |
| Uranium-235 | Lead-207 | 704 million years |
| Thorium-232 | Lead-208 | 14.0 billion years |
| Rubidium-87 | Strontium-87 | 48.8 billion years |
| Potassium-40 | Argon-40 | 1.25 billion years |
| Samarium-147 | Neodymium-143 | 106 billion years |

living plants and animals. In effect, all carbon in living organisms contains a constant proportion of radiocarbon to nonradioactive carbon. After the death of the organism, the amount of radiocarbon gradually decreases as it reverts to nitrogen-14 by radioactive decay. By measuring the amount of radioactivity remaining in organic materials, the amount of carbon-14 in the material can be calculated and the time of death can be determined. For example, if carbon from a sample of wood is found to contain only half as much carbon-14 as that from a living plant, the estimated age of the old wood would be 5,730 years (USGS 2001).

## Did You Know?

The radiocarbon clock has become an extremely useful and efficient tool in dating the important episodes in the recent prehistory and history of man, but because of the relatively short half-life of carbon-14, the clock can be used for dating events that have taken place only within the past 50,000 years (USGS, 2001).

Table 14.3 lists a group of rocks and materials that have been dated by various atomic clock methods:

# Chapter Review Questions

14.1    _____ gives us a sequence of events but not how old something is.

14.2    _____ tells us how old something is.

14.3    _____ deals with the definition and interpretation of strata and stratified rocks.

14.4    _____ states that the geologic processes of today were the same in the past.

14.5    The law of _____ states that all sedimentary rocks are originally deposited horizontally or nearly horizontally.

14.6    This law states that sedimentary rocks are laterally continuous over large geographic areas of Earth: _____.

14.7    _____ posits that the oldest layer of sediment is at the bottom of the sequence, and the youngest at the top.

14.8    An intrusion is always _____ than the rock it invades.

14.9    _____ is known as parallel unconformities.

14.10   The _____ depicts snapshots of geologic time.

14.11   The _____ is the most recent and accurate method yet devised for measuring absolute geologic time.

14.12   _____ is produced continuously in the earth's upper atmosphere.

**Table 14.3.   Rock Groups Dated by Atomic Clock Methods (USGS, 2001)**

| Rock and Material Samples | Approximate Age in Years |
|---|---|
| **Charcoal** | |
| Sample, recovered from bed of ash near Crater Lake, Oregon, is from a tree burned in the violent eruption of Mount Mazama, which created Crater Lake. This eruption blanketed several states with ash, providing geologists with an excellent time zone. | 6,640 |
| **Charcoal** | |
| Sample collected from the ``Marmes Man'' site in southeastern Washington. This rock shelter is believed to be among the oldest known inhabited sites in North America. | 10,130 |
| **Spruce wood** | |
| Sample from the Two Creeks forest bed near Milwaukee, Wisconsin, dates one of the last advances of the continental ice sheet into the United States. | 11,640 |
| **Bishop Tuff** | |
| Samples collected from volcanic ash and pumice that overlie glacial debris in Owens Valley, California. This volcanic episode provides an important reference datum in the glacial history of North America | 700,000 |
| **Volcanic ash** | |
| Samples collected from strata in Olduvai Gorge, East Africa, which sandwich the fossil remains of Zinjanthropus and Homo habilis—possible precursors of modern man. | 1,750,000 |
| **Monzonite** | |
| Samples of copper-bearing rock from vast open-pit mine at Bingham Canyon, Utah. | 37,500,000 |
| **Quartz monzonite** | |
| Samples collected from Half Dome, Yosemite National Park, California. | 80,000,000 |
| **Conway Granite** | |
| Samples collected from Redstone Quarry in the White Mountains of New Hampshire. | 180,000,000 |
| **Rhyolite** | |
| Samples collected from Mount Rogers, the highest point in Virginia. | 820,000,000 |
| **Pikes Peak Granite** | |
| Samples collected on top of Pikes Peak, Colorado. | 1,030,000,000 |
| **Gneiss** | |
| Samples from outcrops in the Karelian area of eastern Finland are believed to represent the oldest rocks in the Baltic region. | 2,700,000,000 |
| **The Old Granite** | |
| Samples from outcrops in the Transvaal, South Africa. These rocks intrude even older rocks that have been dated. | 3,200,000,000 |
| **Morton Gneiss** | |
| Samples from outcrops in southwestern Minnesota are believed to represent some of the oldest rocks in North America. | 3,600,000,000 |

# References and Recommended Reading

Blatt, H., Berry, W. B. N., and Brande, S. 1991. *Principles of Stratigraphic Analysis*. Boston: Blackwell Scientific Publications.

Gould, S. J. 1989. *Wonderful Life*. New York: W. W. Norton & Co.

Harris, E. C. 1979. *Principles of Archaeological Stratigraphy*. New York: Academic Press.

Holmes, A. 1937. *The Age of Earth*. London: Nelson.

Levin, H. L. 2003. *The Earth Through Time*, 7th ed. Hoboken, NJ: Wiley.

Levin, H. L. 2005. *The Earth Through Time*, 8th ed. Hoboken, NJ: Wiley.

Palmer, D. 2005. *Earth Time: Exploring the Deep Past from Victorian England to the Grand Canyon*. New York: Wiley.

Tarbuck, E. J., Lutgens F. K., and Tasa, D. 2007. *Earth: An Introduction to Physical Geology*, 9th ed. Upper Saddle River, NJ: Prentice-Hall.

USGS. 2001. *Radiometric Time Scale*. Washington, DC. Accessed 10/20/08 at http://pubs.usgs.gov/gip/geotime/radiometric.html.

USGS. 2007. *U.S. Geological Survey Fact Sheet 2007-3015: U.S. Geological Survey Geologic Names Committee*, 2007. Washington, DC: USGS.

USGS. 2008. *The Geologic Time Scale*. Accessed 10/10/08 at http://vulcan.wr.usgs.gov/Glossary/geo_time_scale.html.

# Answers to Chapter Review Questions

## CHAPTER 1

1.1   35
1.2   7
1.3   Moho layer
1.4   Flow
1.5   Observe and define the problem
1.6   Parsimony
1.7   Uniformitarianism
1.8   Water cycle
1.9   Lithosphere
1.10  Misplaced soil

## CHAPTER 2

2.1   anything that occupies space and has weight
2.2   intensive
2.3   density
2.4   specific gravity
2.5   luster
2.6   atoms
2.7   molecule
2.8   atomic mass
2.9   bonds
2.10  cation

## CHAPTER 3

3.1   magma
3.2   basaltic, andesitic, rhyolitic
3.3   gabbro
3.4   felsite
3.5   cider cone

3.6   lava spillways
3.7   intrusive rocks
3.8   sills
3.9   stocks or plutons
3.10  fumarole

## CHAPTER 4

4.1   detritus
4.2   clastic
4.3   diagenesis
4.4   conglomerates
4.5   biochemical sediments
4.6   limestone
4.7   evaporites
4.8   lignite
4.9   cross-bedding
4.10  concretions

## CHAPTER 5

5.1   pressure, temperature
5.2   calcite
5.3   facies
5.4   tensional
5.5   shear
5.6   horsts
5.7   slikensides
5.8   folds
5.9   chevron
5.10  mountain building

## CHAPTER 6

6.1   rock type; structure; slope; climate; animals; time
6.2   physical weathering
6.3   chemical weathering
6.4   frost wedging
6.5   carbonic acid
6.6   lichens

## CHAPTER 7

7.1   stream channel
7.2   evapotranspiration
7.3   infiltration capacity
7.4   thalweg
7.5   sinuosity
7.6   meander
7.7   balance
7.8   laminar
7.9   wash load; coarser materials
7.10  aquifers

## CHAPTER 8

8.1   mountain
8.2   fjord
8.3   basal sliding
8.4   cirques
8.5   tarn
8.6   arêtes
8.7   drumlins
8.8   till
8.9   erratics
8.10  eskers

## CHAPTER 9

9.1   transporter
9.2   saltation
9.3   deflation
9.4   ventifact
9.5   eolian
9.6   Barchan dunes
9.7   U-
9.8   8loess
9.9   mass wasting
9.10  slumps

## CHAPTER 10

10.1   salinity
10.2   trench

# INDEX

# About the Author

**Frank R. Spellman** is assistant professor of environmental health at Old Dominion University in Norfolk, Virginia. He holds a BA in public administration, a BS in business management, an MBA, and an MS and a PhD in environmental engineering.

Spellman consults on homeland security vulnerability assessments (VAs) for critical infrastructure, including water/wastewater facilities nationwide. He also lectures on homeland security and health and safety topics throughout the country and teaches water/wastewater operator short courses at Virginia Tech.

Spellman has been cited in more than 400 publications, and he is a contributing author for *The Engineering Handbook*, 2nd edition. He has written more than fifty books that cover topics in all areas of environmental science and occupational health, including *Chemistry for Nonchemists* (2006), *Biology for Nonbiologists* (2007), *Ecology for Nonecologists* (2008), and *Physics for Nonphysicists* (2009), all published by Government Institutes.